SEWAGE TREATMENT
BASIC PRINCIPLES AND TRENDS

SEWAGE TREATMENT
BASIC PRINCIPLES AND TRENDS

R. L. BOLTON, A.M.C.T., L.R.I.C., F.Inst.W.P.C.(Dip.), F.I.P.H.E., F.R.S.H.
Manager, Rochdale Corporation Sewage Purification Department

L. KLEIN, M.Sc., Ph.D.(Lond.), F.R.I.C., Hon.F.Inst.W.P.C.

 ann arbor science publishers inc.
DRAWER NO. 1425 · ANN ARBOR, MICHIGAN 48106

ANN ARBOR SCIENCE PUBLISHERS, INC.
P.O. BOX NO. 1425
ANN ARBOR, MICHIGAN 48106 USA

Library of Congress Catalog Card No. 74–162939

First published in England by
The Butterworth Group (Publishers) Ltd.
All rights reserved.

First published	1961
second impression	1969
Second edition	1971
second impression	1973

© Butterworth & Co (Publishers) Ltd., 1971

ISBN 250 97517 3

Filmset by Filmtype Services Ltd., Scarborough
Printed in England by photo-lithography
and made in Great Britain at the
Pitman Press, Bath

PREFACE TO SECOND EDITION

Since the publication of the First Edition in 1961 there have been many changes in the field of Water Pollution Control. New legislation has come into force which has allowed greater control to be exercised over discharges of effluents into inland, estuarial and tidal waters, and also into underground strata. Additional powers have also been given to sewage authorities to control the discharges of trade waste liquors into municipal sewerage systems, and many advances have been made in the techniques of sewage and trade effluent treatment and sludge disposal. Account has, therefore, been taken of these changed conditions in writing this 2nd Edition. Two additional chapters have been included, one dealing with the need for flow measurement in the treatment of sewage and the general principles involved, and the other gives examples of the types of chemical and associated calculations connected with sewage treatment. The latter chapter is intended primarily as an aid to those students who have not had a formal training in chemical technology. Advantage has been taken to introduce SI units wherever possible.

We would like to express our grateful thanks to Miss Susan Lister for the preparation of some of the additional drawings, and also to the Editor of the Journal *Water Pollution Control* (formerly the *Journal of the Institute of Sewage Purification*) for permission to reproduce in Chapter Seven the drawing of the Banks upward-flow pebble clarifier, and also in Chapter 13 of the problem dealing with Phelps' Law.

1971   R.L.B.
                                                                               L.K.

PREFACE TO FIRST EDITION

IN writing this book our object has been to provide one which will cater for the needs of the person who wishes to study the basic principles of the processes of sewage treatment and to provide a framework on which can be built a comprehensive knowledge of the subject. No attempt has been made to deal in detail with the design and construction of sewage treatment plants, as many excellent books have been written which deal fully with this aspect of the subject. We realise that this book has a bias towards the chemistry of sewage treatment for which we make no apology, as this aspect has quite often been relegated to a minor role. Those we had in mind when writing the book were foremen, working managers and young assistants on sewage works, public health inspectors and students of civil, municipal and sanitary engineering. In the last chapter we have tried to outline some of the trends which are taking place in the field of sewage treatment, but we are fully aware that the list is not comprehensive as this is not possible in a book of this size.

Our grateful thanks are due to the following firms for their kindness in allowing us to reproduce some of their drawings: Ames Crosta Mills and Co. Ltd., Adams Hydraulics Ltd., Dorr-Oliver Co. Ltd., Whitehead & Poole Ltd., and Jones Attwood Ltd.

We wish to express our gratitude to Dr. T. Stones, Manager of the Salford Corporation Sewage Works, who has read through most of the chapters in manuscript and has offered a great deal of valuable advice and constructive criticism; to Mr. R. E. Woodward, Clerk to the Mersey River Board, for his advice on some of the legal sections; and to Mr. H. J. Greenhouse, the Senior Pollution Inspector of the Mersey River Board, who has allowed us to draw freely on his great experience, gained during 44 years' service in the field of river pollution prevention, and for his help and advice at all times.

Our thanks are also due to Mr. J. M. Gaskell for his kindness in preparing some of the drawings; to Mr. R. Kershaw, Manager of the Bury Corporation Sewage Works, for supplying information; and to Mr. J. B. Allcroft, Mrs. S. Rawson, Mr. A. Molyneux and Mr. C. J. Carroll (Laboratory Staff) for reading and criticising manuscripts and proofs.

R.L.B.
L.K.

CONTENTS

1	INTRODUCTION	1
2	NATURE OF SEWAGE AND CHEMICAL ANALYSIS	9
3	SEWERAGE SYSTEMS, STORM SEWAGE OVERFLOWS AND STORM SEWAGE TREATMENT	38
4	PRELIMINARY TREATMENT	48
5	PRIMARY SEDIMENTATION	60
6	AEROBIC BIOLOGICAL TREATMENT	79
7	METHODS OF IMPROVING FINAL EFFLUENTS	109
8	SLUDGE TREATMENT AND DISPOSAL	120
9	FLOW MEASUREMENT	145
10	EFFECTS OF TRADE WASTES	153
11	SMALL SEWAGE TREATMENT PLANTS	186
12	TRENDS IN THE FIELD OF WATER POLLUTION CONTROL	193
13	CHEMICAL CALCULATIONS	220
	APPENDIX 1 *Suggestions for Further Reading*	229
	APPENDIX 2 *British: Metric Units—Conversion Tables*	232
	Index	247

Chapter 1

INTRODUCTION

Sewage can be regarded as the water-borne waste products of man. When the human race was thinly scattered over the face of the earth and tended to be nomadic by nature, the disposal of waste matter did not constitute a serious problem, as large accumulations did not occur and the soil was able to break down the small concentrations into harmless end products. Difficulties arose when man started to congregate in large communities such as London, with a consequent build-up of filth in a restricted area. This resulted in the prevalence of disease epidemics and unsatisfactory living conditions. The trouble was later aggravated by the coming of the Industrial Revolution, with its attendant increase in population and growth of cities and towns. The introduction of the water carriage system in the early 19th Century had the effect of transferring the filth from the streets to the rivers; in certain areas there was further pollution by discharges of trade waste from industrial premises built along the river banks so as to obtain easily accessible water supplies. It became apparent that if man was to live healthily in large communities, it was essential that he should have both a reliable supply of pure water and efficient and hygienic means of disposing of waste products. Towards the middle of the 19th Century the Government became aware of the seriousness of the problem and realised that the watercourses were rapidly becoming open sewers. Various commissions were set up to study and report on the problem, and finally in 1898 a Royal Commission on Sewage Disposal was appointed to investigate and consider methods for the treatment and disposal of sewage and trade wastes. During the following 17 years, 9 valuable Reports and a Final Report[1] summarising the earlier ones were published, which have formed the basis of the science of sewage purification in this and other countries.

The usual means of disposing of sewage is to discharge it into tidal waters or an inland watercourse. Primarily, however, this book will be concerned with the methods of treating sewage to render it fit to be discharged into inland streams.

Clean natural water in a watercourse is normally well oxygenated and contains a large and varied number of life forms, including protozoa, bacteria and aquatic plants and animals, which are interdependent, forming a complex system and keeping the stream in a healthy condition. The water obtains its dissolved oxygen mainly from the atmosphere by surface aeration, and this process is assisted by the turbulence of the stream caused by the velocity of its flow and its passage over rocks, stones and weirs. At 15°C pure water in equilibrium with air contains 10·1 mg/l of dissolved oxygen, but this saturation level varies with the temperature of the water, falling when the temperature rises and rising when the temperature falls. Increase in the salinity of the water at any given temperature reduces its saturation level. *Table 1* gives the saturation levels of clean water at different temperatures, based on recent work of the Water Pollution Research Laboratory.*

The organisms utilising the dissolved oxygen break down organic matter entering the stream, and the salts and carbon dioxide formed are used by the plants, by photosynthetic means, to build up their own structures. It will be seen, therefore, that due to the presence of various organisms and dissolved oxygen, a stream has considerable powers of self-purification, but there is a limit to these powers. If the dissolved oxygen content is seriously reduced, some of the organisms die or are driven away, with the result that, instead of organic matter being oxidised in the normal way, anaerobic organisms begin to build up and break down the organic matter into offensive products, until ultimately a septic condition is set up in the stream water, and foul odours, due to hydrogen sulphide, etc. are emitted.

Let us consider, therefore, what happens if untreated sewage, which contains organic matter in suspension, in colloidal dispersion and in true solution, is discharged into a watercourse. The suspended solids tend to settle out in the slack water and behind weirs, forming banks of sludge. The organic portion of the sludge tends to denude the stream of dissolved oxygen and finally undergoes putrefaction, especially in warm weather, resulting in portions of the sludge being buoyed to the surface by the gas produced. In addition, the inorganic portion of the sludge can help to destroy animal and plant life in the stream, and the organic matter in colloidal dispersion and true solution also plays its part in reducing the dissolved oxygen content.

---

* Earlier results from this laboratory were up to about 3–4 per cent lower due to losses of iodine vapour in the Winkler method of analysis using alkaline potassium iodide. In the later results, a special alkaline sodium iodide reagent was used to avoid these losses.

Introduction

Table 1. SOLUBILITY OF OXYGEN IN PURE WATER AT EQUILIBRIUM WITH AIR SATURATED WITH WATER VAPOUR AT A PRESSURE OF 101 300 N/m² (760 mm Hg)

Temperature °C	Dissolved oxygen mg/l
0	14·6
1	14·2
2	13·8
3	13·5
4	13·1
5	12·8
6	12·4
7	12·1
8	11·8
9	11·6
10	11·3
11	11·0
12	10·8
13	10·5
14	10·3
15	10·1
16	9·9
17	9·7
18	9·5
19	9·3
20	9·1
21	8·9
22	8·7
23	8·6
24	8·4
25	8·3
26	8·1
27	8·0
28	7·8
29	7·7
30	7·6
31	7·4
32	7·3
33	7·2
34	7·1
35	7·0

When sewage is discharged into a watercourse, the dissolved oxygen content of the stream decreases considerably in a short time, but if the stream is in a healthy state the dissolved oxygen will then start to increase, albeit at a slower rate, owing to the breakdown of the organic matter into harmless end products and to re-aeration of the water (see *Figure 1*). If, however, the volume of sewage is large in relation to the volume of stream water and strong in character, its oxygen demand will be greater than the dissolved oxygen present in the stream and that supplied by re-aeration, with the result that the stream will be reduced to a polluted condition and may eventually become septic.

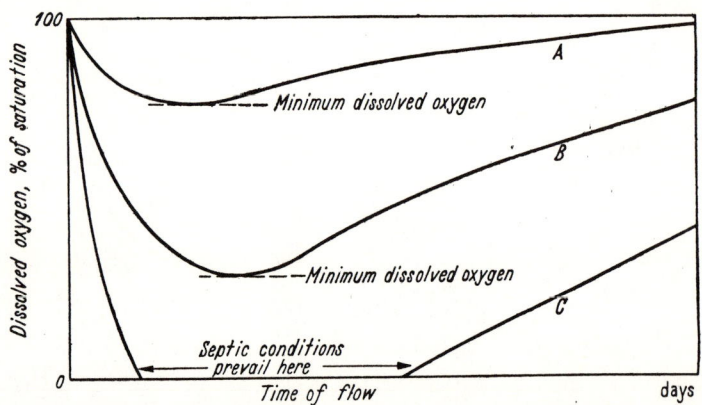

Figure 1. Oxygen sag curves showing degradation and recovery of a polluted stream after (A) slight, (B) heavy and (C) gross pollution

The Royal Commissioners[1] realised that the question of dilution of the sewage was of prime importance, and in Volume I of their 8th Report suggested the following standards for sewage effluents, based on the degree of dilution they would receive on discharge:

Above 500 dilutions   Crude sewage may be discharged after screening and if necessary detritus tank treatment.
300–500 dilutions   Suspended solids not to exceed 150 mg/l.
150–300 dilutions   Suspended solids not to exceed 60 mg/l.
8–150 dilutions   Five days' Biochemical Oxygen Demand (18·3°C)* not to exceed 20 mg/l and suspended solids not to exceed 30 mg/l (the normal 'Royal Commission Standards' for sewage effluents discharging to inland streams).

They also recommended in Volume II of their 8th Report that conditions should be such that a stream which has received discharges of effluent should have a B.O.D. not exceeding 4 mg/l, and the dissolved oxygen should not be below 4 ml/l at 18·3°C (presumed summer maximum temperature). The view was taken that if these limits were not infringed, the stream would not be likely to cause nuisance or show any appreciable signs of pollution. It should be fully

---

* Nowadays 20°C

Introduction

understood though that the Commissioners assumed that the water in the stream into which an effluent was discharged would be clean water. They suggested (8th Report, Volume I) that stream water should be classified on the basis of its five days' Biochemical Oxygen Demand in the following manner:

Classification	Five days' B.O.D.
Very Clean	1 mg/l
Clean	2 mg/l
Fairly Clean	3 mg/l
Doubtful	5 mg/l
Bad	10 mg/l

The B.O.D. of the mixture of sewage effluent plus river water can be calculated from the expression $(x + yz)/(z + 1)$ given by the Royal Commission where

x = B.O.D. of effluent (mg/l)
y = B.O.D. of river water above outfall (mg/l)
z = dilution factor (proportion of river water to effluent)

It can be seen from the above that if a sewage effluent with a B.O.D. of 20 mg/l is discharged into a stream where it receives a dilution of 8:1 with clean water (B.O.D. 2 mg/l), the stream below the outfall will have a 5 days' B.O.D. which will not exceed 4 mg/l.

The standards of 20 mg/l B.O.D. and 30 mg/l suspended solids for sewage effluents have largely become accepted in this country. It will be seen, however, that if (a) dilution in the receiving stream is less than 8:1, or (b) the stream water is not clean or (c) downstream from the outfall a particularly high degree of purity of the stream water is required (e.g. for drinking water supply), there are good grounds for requiring the effluent to comply with standards stricter than those recommended by the Royal Commissioners. It must also be stressed that the standards are maximum ones, and it should be the aim of every sewage works manager to ensure that his effluent is at all times *well within* these standards and is the best his plant is capable of producing.

While it may be a sound economic principle to see that expenditure on new or reconstructed sewage works does not go beyond the point necessary to ensure that the effluent can attain the required standards at all times, it is false economy to construct a works whose capacity is only sufficient to deal with immediate requirements; capacity should be adequate for dealing with any probable increase in the volume of sewage to be treated. Consideration should also be given to long term increases in the volume of sewage and the works should be so designed that additional units can be constructed in the easiest and most economical manner and with the minimum of interference with the operation of the existing units.

It has been stated many times that no two sewages are exactly alike, which is one of the factors making the science of sewage purification so interesting, and works have to be designed to meet the specific characteristics of the sewage to be treated. Nevertheless, the processes in all sewage works basically fall more or less into a general pattern:

(a) Removal and/or disintegration of gross solid matter
(b) Removal of grit
(c) Separation and treatment of storm sewage
(d) Removal of suspended solids
(e) Aerobic biological treatment
(f) Removal of suspended solids from biologically oxidised sewage
(g) Final improvement of effluent
(h) Sludge treatment and disposal

The methods of carrying out the various stages of treatment vary from works to works, as may also the sequence, and some stages may be omitted altogether.

The choice of a site for a sewage works is a very important matter, and careful consideration has to be given to many factors. The site must obviously be within a reasonable distance from the watercourse into which it is proposed to discharge the effluent. Wherever possible the plant should be sited at the lowest point of the drainage system, the outfall should be above the flood level of the watercourse, and if possible there should be sufficient fall through the site (the amount being dictated by the type of plant used) to obviate pumping, thus reducing running costs. The ground on the site should be sound, so as not to require constructing costly foundations, and should be free from flooding. There should also be sufficient land available for future extensions, let alone for immediate requirements. The site should preferably be at a reasonable distance from built-up, especially residential, areas, easily accessible from main roads and within reasonable walking distance of a public transport route—a point often overlooked that may cause difficulties in obtaining labour. It will be realised that all these conditions can rarely be satisfied, e.g. in many cases pumping will have to be carried out either on the sewerage system or at the works. However, all these points should be considered when choosing a site.

Where a new discharge of sewage effluent is to be made to a stream or a new outlet is to be constructed, application has to be made to the

River Authority having jurisdiction over the stream for their consent, in accordance with the provisions of Section 7 of the Rivers (Prevention of Pollution) Act, 1951[2]. The River Authority may, in the case of new discharges, impose conditions regarding the nature and composition, temperature and volume or rate of discharge of the effluent. In the case of sewage effluents the conditions regarding nature and composition of the effluent are usually based on the Royal Commission's recommended standards for such effluents, but where stream conditions warrant it River Authorities are now tending to impose more stringent conditions. In the case of new outlets, conditions can be imposed regarding the point of discharge, construction of the outlet and the use to which it will be put. River Authorities' consents to new outlets are not required if their construction or alteration, or the raising of a loan to defray their costs, have been approved by the Minister of Housing and Local Government.

It should be remembered that alterations to existing discharges or outlets will make them 'new' discharges under the provision of Section 7 of the 1951 Act.

Appeal can be made to the Minister of Housing and Local Government if it is considered that a River Authority has unreasonably withheld consent or has imposed unreasonable conditions.

New or altered discharges of trade or sewage effluent and new or altered outlets to specified estuarial and tidal waters were brought under the same control as those to inland streams by the Clean Rivers (Estuaries and Tidal Waters) Act, 1960[3]. This Act applies the provisions of Section 7 of the Rivers (Prevention of Pollution Act, 1951, subsections 1–15) to new and altered discharges and outlets to those estuarial and tidal waters whose limits are specified in the Schedule to the 1960 Act, which covers 95 estuaries in England and Wales.

The Rivers (Prevention of Pollution) Act, 1961[4], made it obligatory for those persons responsible for making a discharge to a stream of trade or sewage effluent, for which a consent had not been obtained under the provisions of Section 7 of the 1951 Act, to make application to the appropriate River Authority, before the 1st June 1963, for a consent to continue making that discharge. The River Authority can withold consent, or issue a consent which may contain conditions with respect to the nature and composition, temperature and volume or rate of discharge. There is no limit on the time a River Authority may take to determine an application, but the discharger is protected until his application is determined providing his effluent complies with the information given in his application, and that he operates his treatment plant as efficiently as at the time of making the application. A consent issued by a River Authority under the provisions of the 1961 Act comes into force 3 months after the date of issue. An appeal can be made to the Minister of Housing and Local Government if a discharger considers that a consent has been refused unreasonably or if he thinks that the conditions of a consent are unreasonable.

It will be seen, therefore, that when all applications, made under the provisions of the 1961 Act, have been determined, all existing and new discharges of trade and sewage effluent to inland streams will be controlled by the conditions of consents issued by the River Authorities. In the same way all new or altered discharges to scheduled estuarial and tidal waters, made since the coming into force of the 1960 Act, will also be controlled by the terms of the consents issued by the River Authorities.

The Water Resources Act, 1963[5] also gave River Authorities, in England and Wales, the power to control, by means of consents, discharges of trade and sewage effluent into underground strata.

In Scotland, River Purification Boards having control over pollution prevention only*, take the place of the English and Welsh all-purpose River Authorities, and the principal Act, the Rivers (Prevention of Pollution) (Scotland) Act, 1951[6], is analogous in many respects to the 1951 Act applying to England and Wales. The Scottish Act of 1951 controlled all new discharges under Section 28, which thus corresponds with Section 7 of the 1951 Act applying to England and Wales. But discharges of effluent which existed before the 1951 Scottish Act came into force, and were not controlled as new discharges by Section 28 of that Act, are now brought under control by Section 1 of the Rivers (Prevention of Pollution) (Scotland) Act, 1965[7], and so require the River Purification Board's consent. Also certain provisions of the 1951 and 1965 Scottish Acts are now applied to tidal (or 'controlled') waters. Broadly speaking, therefore, the 1965 Scottish Act has the effect of combining the powers which the English and Welsh River Authorities have under the Clean Rivers (Estuaries and Tidal Waters) Act 1960 and the Rivers (Prevention of Pollution) Act, 1961.

The Water Resources Act, 1963, does not apply to Scotland, where conditions, especially higher rainfall, ample water resources, and less pollution generally, are very different from those in England.

REFERENCES

1. *Royal Commission on Sewage Disposal*, 1901–15, 9 *Reports* with Numerous Appendices; also a *Final Report*, 1915. London; H.M.S.O.
2. *Rivers (Prevention of Pollution) Act*, 1951. London; H.M.S.O.
3. *Clean Rivers (Estuaries and Tidal Waters) Act*, 1960. London; H.M.S.O.
4. *Rivers (Prevention of Pollution) Act*, 1961. London; H.M.S.O.
5. *Water Resources Act*, 1963. London; H.M.S.O.
6. *Rivers (Prevention of Pollution) (Scotland) Act*, 1951. London; H.M.S.O.
7. *Rivers (Prevention of Pollution) (Scotland) Act*, 1965. London; H.M.S.O.

* Unlike their English and Welsh counterparts, they are not concerned with land drainage or with fisheries.

Chapter 2

NATURE OF SEWAGE AND CHEMICAL ANALYSIS

NATURE OF SEWAGE

Domestic sewage consists of discharges of spent water from wash-basins, bathrooms and washing machines (soapy and dirty water), kitchens (food materials, dirty water) and lavatories (urine, faeces, paper). It is a complex mixture of mineral and organic matter in many forms, including (a) large and small particles of solid matter floating and in suspension, (b) colloidal and pseudo-colloidal dispersion and (c) true solution*. Sewage also contains living matter, especially bacteria, viruses and protozoa; it is an excellent medium for the development of bacteria, some of which may be pathogenic (disease-producing), and in the crude state contains many millions of bacteria per millilitre. Ground water leaking into the sewers and surface water from streets, roofs, etc. may also be present in sewage. Most towns and cities in this country also contribute to sewage a variety of industrial waste waters, e.g. from textile mills, chemical works, gas works, tanneries, slaughterhouses, pulp and paper mills, etc.

The water content of sewage is very high (99·9 per cent or more), which means that the total solid matter is only 0·1 per cent or less. The solid portion contains paper fibres (cellulose), matches, food materials, soaps, fats, oils, greases and faeces as well as insoluble mineral matter (sand, clay, gravel, etc. washed from the streets by

---

* As a rough guide, particle sizes are approximately:-
suspensions (i.e. relatively coarse particles which settle rapidly)   >500 nm
colloidal solutions (very fine particles which do not settle)   1–500 nm
true solutions (ions and molecules which do not settle)   <1 nm
(The abbreviation 'nm', or nanometre (= 10^{-9} m), is now preferred internationally to the older 'mμ'.)
No sharp demarcation line exists between these three groups: e.g. there is a borderline 'pseudo-colloidal' region (particle size around 500 nm) with intermediate settling characteristics between colloidal solutions and suspensions.

rain). Amongst the organic substances present in sewage are carbohydrates, lignins (complex compounds of carbon, hydrogen and oxygen present in wood), fats, soaps (metallic salts of fatty acids), synthetic detergents and proteins and their decomposition products. Ammonia and ammonium salts are always present, being produced by the decomposition of complex nitrogenous organic matter. The objectionable character of sewage is due mainly to the presence of nitrogenous, sulphur-containing and phosphorus-containing organic matter which readily undergoes putrefaction by anaerobic bacteria, with formation of foul-smelling compounds; these include malodorous hydrogen sulphide, organic sulphides and mercaptans, and also certain organic amines (especially indole and skatole) which impart a characteristic unpleasant faecal odour. The sulphur compounds in sewage include not only proteins and their decomposition products but also synthetic detergents (mainly organic sulphonates) and inorganic sulphates.

Some typical examples of a few organic substances in domestic sewage are given in *Table 2*.

Urine, which is present in sewage, contains about 1 per cent sodium chloride (NaCl), about 2·5 per cent urea and various complicated organic substances. Many other substances in sewage derive from trade wastes (phenols, sulphides, formaldehyde, cyanides, metallic contaminants, free chlorine, etc.). Dissolved mineral salts in sewage (such as chlorides, sulphates, phosphates* and bicarbonates of sodium, potassium, ammonium, calcium, magnesium and iron) come from the water supply, from urine and from any trade wastes present. It has been shown that domestic sewage contains traces of metals such as zinc, copper, chromium, manganese, nickel and lead.

When the domestic water supply is hard (i.e. high in calcium and magnesium content), insoluble calcium and magnesium soaps are precipitated during washing; these eventually settle out with other organic material as sludge when the sewage is settled in sedimentation tanks. For instance, sodium stearate (a sodium soap) interacts with dissolved calcium sulphate according to the equation

$$2C_{17}H_{35}CO_2Na + CaSO_4 = (C_{17}H_{35}CO_2)_2 Ca + Na_2SO_4$$
$$\text{sodium stearate} \qquad\qquad\qquad \text{calcium stearate}$$
$$\text{(soluble)} \qquad\qquad\qquad\qquad \text{(insoluble)}$$

Sewage arriving at a sewage works may be fresh, i.e. will contain some dissolved oxygen and sometimes small amounts of nitrite and nitrate, especially after rain. Stale sewage usually has no dissolved

---

* Part of the phosphate in sewage is derived from 'builders' contained in commercial synthetic detergent preparations.

Nature of Sewage and Chemical Analysis

Table 2. TYPICAL ORGANIC MATERIALS IN DOMESTIC SEWAGE

Carbohydrates
  cellulose and starch $(C_6H_{10}O_5)n$
  cane sugar and lactose $C_{12}H_{22}O_{11}$
  glucose $C_6H_{12}O_6$
  pentose sugars $C_5H_{10}O_5$

Fats
  General formula
  $$\begin{array}{l} CH_2OR' \\ | \\ CHOR'' \\ | \\ CH_2OR''' \end{array}$$
  (where R', R'' and R''' are different or identical fatty acid residues)

Proteins

⋮NH—CHX'—CO⋮NH—CHX''—CO⋮NH—CHX'''—CO⋮etc.

Skeleton of protein molecule

repeated indefinitely, often more than 1000 times, made up of amino-acid residues condensed together with elimination of water. X', X'', X''', etc. = organic groups. (Sulphur and phosphorus can also be present in the molecule.)

Protein decomposition products
  urea $NH_2.CO.NH_2$
  glycine (the simplest amino-acid) $CH_2(NH_2).COOH$
  cysteine (a sulphur containing amino-acid) $CH_2(SH).CH(NH_2).COOH$

  indole
  $$H_4C_6 \underset{NH}{\overset{CH}{\diagup\diagdown}} CH$$

  skatole
  $$H_4C_6 \underset{NH}{\overset{C(CH_3)}{\diagup\diagdown}} CH$$

  fatty acids (R = alkyl group) $R.COOH$

Synthetic detergents*
  A typical anionic synthetic detergent is
  sodium alkylbenzene sulphonate $R.C_6H_4.SO_3Na$
  (R = long hydrocarbon chain)

---

* Sewage in this country contains on the average about 15 mg/l of anionic synthetic detergents (expressed as Manoxol O.T., or sodium dioctylsulphosuccinate). These detergents were formerly of the 'hard' type (resistant to biochemical degradation) but, since the beginning of 1965, have been replaced by a 'softer' type more easily amenable to biochemical oxidation. (see Chapter 12)

---

oxygen. In septic sewage, the development of anaerobic bacteria will have caused septic action, characterised by blackening and production of foul-smelling sulphuretted hydrogen (H_2S).

Sulphuretted hydrogen and sulphides are produced by the anaerobic fermentation of organic sulphur compounds and by bacterial reduction of sulphates. In very septic sewage, as much as 60 mg/l of

H₂S has been reported. The amount of this substance will depend upon the strength, age, temperature, redox potential and sulphate content of the sewage and also on the content of proteolytic (protein-splitting) and sulphate-reducing bacteria present. In aqueous solution, H₂S ionises slightly according to the equation

$$H_2S \rightleftharpoons H^+ + HS' \rightleftharpoons H^+ \; S''$$
(very slight)

Hydrogen sulphide, if present in gaseous form, will betray itself by its very objectionable odour of rotten eggs. The proportion present in molecular form (i.e. liable to cause bad smells) depends upon the pH value and is given in *Table 3*. When present in molecular (unionised form), H₂S tends to escape to the air where it causes nuisance and blackens lead paints and metals.

Table 3. RELATION BETWEEN pH VALUE AND PROPORTION OF SULPHIDE PRESENT AS MOLECULAR UNIONISED H₂S IN SEWAGE

pH	Approximate percentage of S in molecular form (H₂S)
5·0 (acid)	98
6·0	83
6·5	61
7·0 (neutral)	37
7·5	14
8·0	5
8·4	2
9·6 (alkaline)	0·1

It will be seen that the more alkaline the sewage (i.e. the higher the pH), the less H₂S is present in molecular form, and consequently the less likely is aerial nuisance. On the other hand, under acid conditions (pH below 7·0), the greater part of the sulphide is present as H₂S and not as S″ or HS′ ions. H₂S is easily oxidised by bacteria in the presence of air (e.g. at or above the water level in sewers) to sulphuric acid (H₂SO₄), which can cause destruction of concrete sewers and corrosion of metal work and, owing to its poisonous nature, endanger men working in sewers, manholes and pump-wells. Corrosion of sewer concrete is particularly serious when the concentration of sulphide is about 2·0 mg/l. At a pH of 7·5 or higher, the production of H₂S by sulphate-reducing bacteria ceases. Polyvinyl chloride plastics sheets and clay lining blocks have been proposed for protecting the inner surfaces of sewers against the ravages of hydrogen sulphide.

Sulphates in sewage can also cause direct chemical attack on concrete, especially if present to the extent of 1 000 mg/l or more (expressed as SO_3). In such instances special, more expensive sulphate-resisting Portland cement, containing a high Al_2O_3 content, can be used. Where this is not possible, limits may have to be placed in trade effluent agreements on the concentration of sulphate permitted in the discharges to the sewers. Chlorination of sewage (up to about 50 mg/l) appears to be the simplest method of preventing corrosion troubles. In South Africa, addition of lime (70 mg/l of CaO) to render the sewage alkaline has proved particularly effective in preventing sewer corrosion.

AEROBIC AND ANAEROBIC PROCESSES

Many of the most important sewage treatment processes are biological in character, utilising the activities of bacteria and other micro-organisms. These processes can be classified into two groups, aerobic and anaerobic.

AEROBIC PROCESSES

Aerobic processes depend upon the presence of aerobic bacteria which require the presence of free oxygen (air) for their metabolism. It is important, therefore, to ensure that at all times there is an adequate air supply. Examples of such processes are percolating filters, the activated sludge process (see Chapter 6) and land treatment. The reactions taking place in the aerobic oxidation of organic matter can be summarised as follows:

carbon, C → CO_2 +carbonates and bicarbonates
carbon (+some insoluble humus[*])
dioxide

hydrogen, H → H_2O
water

[*] Humus is a dark brown, almost black, complex organic material containing C, H, O and N which is very resistant to microbiological decomposition.

nitrogen, N → NH₃ → HNO₂ → HNO₃
ammonia   nitrous acid   nitric acid
(neutralised by bases present)

sulphur, S → H₂SO₄ (neutralised by bases present)
sulphuric acid

phosphorus, P → H₃PO₄ (neutralised by bases present)
phosphoric acid

The conversion of ammonia to nitrites and nitrates, an important process in normal percolating filters, is generally called 'nitrification'*. The oxidation and stabilisation of sewage in a Biochemical Oxygen Demand test proceeds in two stages. First, the carbonaceous matter is oxidised and this may take up to about 20 days. Then nitrification (i.e. oxidation of NH_3 ultimately to nitrate) sets in and may continue for 2–3 months. There can be some overlapping of these stages and frequently some nitrification takes place much earlier and simultaneously with carbonaceous oxidation.

A typical curve showing the uptake of dissolved oxygen by settled sewage in a respirometer is shown in *Figure 2*.

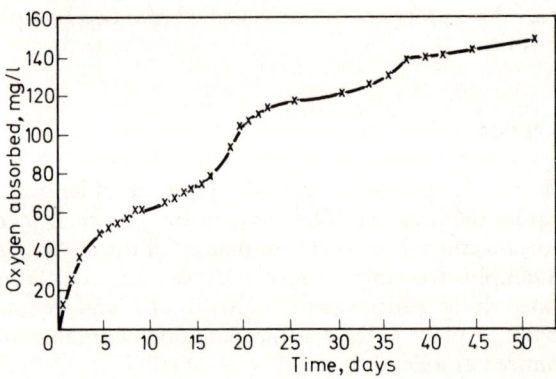

Figure 2. Uptake of dissolved oxygen by settled sewage in a respirometer at 20°C. (By courtesy of the Surveyor *and Dr. B. A. Southgate)*

During the first stage of carbonaceous oxidation the rate of deoxygenation at any one moment is directly proportional to the amount of organic matter remaining and the reaction is a monomolecular one. This is expressed by Phelps' law[2] according to the equation:

$$\log_{10} \frac{L}{L-x} = kt \qquad (1)†$$

* Some authors refer to the conversion of ammonia to nitrite as nitrosification.
† There are other forms of this equation: $\log_{10} \frac{L-x}{L} = -kt$, and $x = L(1-10^{-kt})$.
The form given above [equation (1)] is the simplest form for calculations.

where L = oxygen demand (mg/l) during first stage (carbonaceous) oxidation (a measure of the total carbonaceous impurity)

x = oxygen demand (mg/l) in t days (as determined by B.O.D. test)

k = deoxygenation constant (0·1, for domestic sewage at 20°C, using common logarithms to the base 10 and time units in days. It would have a different value for other units and for other types of waste liquors). It is a measure of the rate of carbonaceous oxidation. The larger the value of k, the more easily does oxidation occur.

At any other temperature, k varies in accordance with the equation:

$$k_T = k[1\cdot047^{(T-20)}] \qquad (2)*$$

where k_T = constant at $T°C$.

The first stage demand is affected by temperature as follows:

$$L_T = L_{20}[1+0\cdot02(T-20)] \qquad (3)$$

The use of these three formulae enables one to calculate the B.O.D. of a sample at different temperatures and times.

These formulae have been much used in stream pollution studies in the deep rivers of the U.S.A., but are difficult to apply to the shallow, much smaller British streams, where there are complications due to the oxygen demands of bottom muds and of organic suspended matter and also, in many instances, to the large number and variety of closely-spaced polluting discharges. Moreover, the incidence of nitrification and photosynthesis adds to the difficulties.

ANAEROBIC PROCESSES

In the absence of free oxygen, such as when sewage turns septic or a river receives a massive pollution by sewage or organic trade waste, anaerobic bacteria† are able to break down complex organic matter. These bacteria obtain their oxygen from substances such as sulphates, phosphates and various organic compounds. Anaerobic processes occur in septic tanks, Imhoff tanks and in separate sludge digestion tanks.

The oxidation and reduction reactions occurring in the anaerobic breakdown of organic matter are as follows:

carbon → organic acids ($R.COOH$) → CH_4 + CO_2 (+some insoluble humus)
           methane carbon dioxide

---

* Other formulae have also been proposed.

† Facultative bacteria (e.g. faecal streptococci) can grow in the presence or absence of free oxygen and thus can act aerobically or anaerobically.

Table 4. IMPORTANT TESTS USED IN SEWAGE ANALYSIS

Test	Performed on	Matter present and/or significance	References
†Permanganate value 3 min	S T F	easily oxidised inorganic and/or organic matter	3, 4
†Permanganate value, 4h	S T F	organic matter: indication of 'strength' of sewage, of waste or tank effluent and of quality of final effluent	3, 4, 5a
Organic carbon	S T F		3–5
*B.O.D., 5 days	S T F		3–5
Dichromate value	S T		4, 5
*Suspended solids	S T F	insoluble organic + mineral solids; sludge-forming material	3–5
†Settleable solids	S T F	solids removed by settlement; sludge-forming material	3–5
Total solids	S T F	total organic + mineral solids; volatile portion = measure of sewage strength	3–5
Dissolved solids	S T F	high value in dry weather, when trade wastes present, or if sewage septic	3–5
Combined nitrogen (ammoniacal, albuminoid or organic, nitrous and nitric)	S T F	relative proportions indicate degree of stabilization of nitrogenous organic matter; nitrates are the end product of purification and the most stable form	3–5, 5a
Chloride	S T F	rough measure of sewage strength: high in dry, low in wet weather; unchanged during sewage purification	3–5
Sulphate	S	may cause corrosion of concrete sewers	3–5
Sulphide	S T	may cause corrosion of concrete and blackening of metalwork and lead paint. Present in septic sewage, sulphide dye, tannery and viscose rayon wastes. Dangerous to men working in sewers	3–5
†Stability	F	liability of effluent to undergo change	3, 5, 5a

Table 4 (continued)

Test	Performed on	Matter present and/or significance	References
†pH value	S T F	intensity of acidity or alkalinity: high or low value indicates trade wastes	3–5
Acidity or alkalinity	S T F	amount of acid or alkali present	3–5
†Transparency	S T F	turbidity due to suspended matter and sewage (or other) colloids	3, 4
Extraction with organic solvent	S T	fats, grease, oils and other immiscible liquids	4, 5, 5a
Anionic syndets	S T F	detergents causing froth or foam	3, 4
Dissolved oxygen	F M	efficiency of aeration in activated-sludge plant	3, 5
‡Metals	S T F	metallic trade wastes	3–5a
‡Cyanide	S T F	electroplating and case-hardening wastes, dangerous to men working in sewers	3–5a
‡Phenols	S T F	gas liquor; synthetic resin wastes	3–5
‡Thiocyanate	S T F	gas liquor	3–5
Iron	S T F	mine water; infiltration water; iron pickle liquor	3–5

S = raw sewage   T = tank effluent   F = final effluent to river   M = mixed liquor (mixture of sewage + activated sludge undergoing aeration)
† Tests suitable for small sewage works with limited laboratory facilities
* Standard tests recommended in the Eighth Report of the Royal Commission on Sewage Disposal, for assessing the quality of final sewage effluents[1]
‡ Toxic substances in trade wastes adversely affecting sewage treatment processes

nitrogen →amino-acids ($R.NH_2.COOH$)→ NH_3 +amines
                                                                    ammonia

sulphur   → H_2S +organic S compounds
                    hydrogen sulphide

phosphorus → PH_3 +organic P compounds
                         phosphine

Thus, it can be seen that anerobic decomposition (or 'putrefaction') gives end products rather different from those of aerobic processes. In particular, methane is produced in large quantities during anaerobic action, and so are a number of compounds with objectionable odours (e.g. H_2S and organic sulphur compounds). In sewage sludge digestion (see Chapter 8) the formation of these, and of volatile organic acids which inhibit digestion, is prevented or minimised by 'seeding' the raw sludge with sufficient well-digested sludge, by maintaining it in a slightly alkaline condition (preferably pH 7·2–7·8) and by avoidance of overloading with raw sludge.

Oxidation and reduction play a vital part in biological sewage treatment processes. A useful yardstick for indicating the intensity of oxidising and reducing conditions in biochemical systems is the oxidation-reduction ('redox') potential, defined as the e.m.f. (in mV) set up in a liquid due to the concentrations of oxidising and reducing substances present. It is measured by immersing a platinum electrode in the sample and determining the potential relative to that of a standard hydrogen (or reference) electrode[*]. In general, aerobic systems (percolating filters, activated-sludge plants) require positive redox potentials of the order $+200$ to $+600$ mV for the most favourable results, as against anaerobic processes which require negative redox potentials (e.g. in sludge digestion plants of the order -100 to -200 mV). The measurement of the redox potential is by no means easy or reliable but although there are technical difficulties, the method is already being used to some extent in the development of automation at some sewage works.

CHEMICAL ANALYSIS OF SEWAGE AND INTERPRETATION OF RESULTS

One of the most important effects of the discharge of sewage and other organic wastes to rivers is a reduction in the concentration of dissolved oxygen in the receiving watercourse owing to its absorption by the organic matter in the presence of bacteria. If the organic pollution load is sufficiently great, the stream may be completely denuded of its dissolved oxygen, thus causing putrefaction and foul odours due to H_2S, etc. (compare anaerobic processes, above). A knowledge of the amount of organic matter present in sewage is,

---

[*] Wastes containing oxidising agents (nitrates, free chlorine) have positive, those containing reducing agents (septic sewage, sulphides) have negative redox potentials.

therefore, of the greatest practical importance. Thus, many of the determinations used in sewage analysis are really tests for the presence of organic matter. Since organic matter, in general, takes up oxygen fairly easily, the amount of oxygen absorbed can be used as a rough measure of the organic content of the sample. Among the most useful of these tests are the purely chemical 'permanganate value' test and the well known biochemical test devised originally by the Royal Commission on Sewage Disposal and now known as the 'Biochemical Oxygen Demand' (B.O.D.) test. Taken in conjunction with the determination of suspended solids, these two tests allow one to form an estimate of the size of plant needed to treat a particular sewage, to determine the performance of the various units and to assess the quality of the final purified effluent. A number of further analytical determinations are also useful, in some cases essential, in supplementing the information provided by the three fundamental tests and in assisting in the control of the sewage works. The most important of the tests are shown in condensed form, with their significance, in *Table 4*. The references in the last column are to the standard and other publications which give practical details of the tests. Not all the tests given in the Table will be used at every works; those specially suitable for small works are indicated. From the point of view of plant control the purification achieved at each stage can be quickly assessed on the basis of a few rapid and simple tests, especially the 4 hours' permanganate value, ammoniacal and albuminoid nitrogen, and the settleable solids (Imhoff cone) tests, without waiting for the B.O.D. and suspended solids tests which take much longer. In *Table 5* are given the more important tests used in sewage sludge analysis and in the control of activated sludge plants.

In the larger laboratories, where expense is not vital, many analyses can be performed quickly and automatically using the Auto Analyser (address: Technicon Instruments Ltd, Chertsey, Surrey). Moreover, many physical and physico-chemical techniques can be applied (e.g. paper, thin-layer, and gas chromatography, spectrophotometry, atomic absorption spectroscopy, polarography, and electrometric titrimetry procedures. These can be used alone or in conjunction with such chemical methods as solvent extraction, ion exchange, complexometric titration and precipitation with collectors or 'scavengers'. The Auto Analyser has been used for the determination of NH_3, NO_2', NO_3', PO_4''', Ca, Mg, heavy metals, borate, silicate and syndets.

Because sewage is continually changing in nature and composition throughout the day owing to fluctuating discharges of trade wastes, the washing and other habits of the population, the effect of storms, etc., it is usual to collect a composite sample for analysis made up of

Table 5. IMPORTANT TESTS USED IN SEWAGE SLUDGE ANALYSIS AND FOR CONTROL OF ACTIVATED-SLUDGE PLANTS

Test	Type of sludge	Significance	References
Moisture; total and volatile solids	R, D, H, A	moisture removal in sludges; solids balance and progress of sludge digestion	3, 5
pH value	D	condition of digester contents	3–5
Alkalinity to methyl orange	D	condition of digester contents	3, 5
Volatile acids	D	early indication of condition of digester contents	5
Extraction with light petroleum	R, D, H, A	grease content: progress of sludge digestion	3, 5
Fertiliser constituents (N, K, P, etc.)	R, D, H, A	manurial value of sludge	6, 7
Buchner funnel: filter leaf; specific resistance	R, D, H, A (especially after chemical conditioning)	sludge filterability and cake yield, on vacuum filter	8–10
Settleability	M	ease of settlement of activated sludge; proportion of activated sludge after 1 h settlement	5, 11
Sludge volume and density indexes	M	conditon or quality of activated sludge	5
Microscopic examination	A, M	condition or quality of activated sludge	11

R = raw sludge (sedimentation tank)   D = digested sludge   H = humus sludge   A = activated sludge
M = mixed liquor from aeration tank of activated sludge plant

samples taken at regular intervals over the day (say, hourly) and, preferably, mixed in proportion to the rates of flow. Automatic samplers are now available commercially for doing this.

SIGNIFICANCE OF THE TESTS AND INTERPRETATION OF RESULTS OF ANALYSIS

PERMANGANATE VALUE (OXYGEN ABSORPTION FROM PERMANGANATE)

This old-established test, performed under strictly standardised conditions, measures the oxygen taken up by the sample from a dilute sulphuric acid solution of N/80 potassium permanganate* when kept in a stoppered bottle for 3 min and 4 h at 27°C. At the end of the incubation period excess of potassium iodide is added; this reacts with the unused permanganate to give iodine which is estimated by titration with standard sodium thiosulphate. If oxidising agents derived from trade wastes are present (e.g. free chlorine and chromate), a low or even 'negative' result may be obtained. In such cases, a blank determination is carried out with the sample+dilute sulphuric acid+potassium iodide to estimate the iodine equivalent of the oxidising agent for correction. If ferric salts are present, syrupy phosphoric acid is added just before the potassium iodide, whereby a complex ferri-phosphate is formed which does not liberate iodine from potassium iodide. If nitrites are present (as in some sewage effluents), they will be included in the figure obtained for the permanganate value. Some workers correct the result so obtained (by deducting the nitrite $N \times 1 \cdot 14$); if so, the fact should be stated. Alternatively, the nitrite can be destroyed by addition of urea to the acidified sample before carrying out the permanganate test.

The 3 min test measures the oxygen demand due to oxidisable inorganic matter (e.g. nitrites, sulphides, sulphites, thiosulphates and ferrous salts) as well as very easily oxidisable organic matter, such as phenols. The 4 h test is often taken as a rough measure of the organic impurity of the sample, allowance being made for any oxidisable inorganic pollution as indicated by the 3 min test.

A relatively high 3 min value thus indicates the presence of much easily oxidised matter and may supply an important clue to the nature of the trade wastes present.

Oxidation during the 4 h test is generally not as complete as in

* In the 4th Royal Commission Report[1] N/8 potassium permanganate was specified, and this stronger permanganate is still used when determining the strength of sewage by the McGowan formula.

the 5-day B.O.D. test. In the 4 h permanganate test, only about 15–20 per cent of the carbonaceous matter of sewage undergoes oxidation, whereas in the 5-day B.O.D. test about 65–70 per cent is oxidised. On the other hand, gas liquor is oxidised by permanganate to the extent of about 70 per cent, but B.O.D. tests give poor oxidation because the toxic constituents of the liquor have an adverse effect on the bacteria.

For a particular sewage or waste, the ratio between the 5-day B.O.D. and the 4 h permanganate value is usually roughly constant; consequently, the permanganate test, which is quick to perform, can be used to give an approximate idea of the B.O.D. Sedimentation of sewage in settling tanks can be expected to reduce the 4 h permanganate value by about 30 per cent; complete biological treatment of sewage generally removes about 75–85 per cent of this value.

A well purified final effluent from an aerobic biological process (percolating filters or activated sludge) should not, in general, exceed about 15 mg/l in permanganate value, but if certain trade wastes are present in the sewage, the permanganate value may be much higher (up to about 25 mg/l), even though the 5-day B.O.D. may not rise above the Royal Commission standard of 20 mg/l.

BIOCHEMICAL OXYGEN DEMAND

This determination, originally put forward by the Royal Commission on Sewage Disposal, is a biochemical test dependent on the activities of bacteria which in the presence of oxygen feed upon, and consume, organic matter. It expresses the amount of oxygen (in mg) taken up by one litre of the sample (usually diluted with sufficient well-oxygenated water) when incubated at a standard temperature (now 20°C) in 5 days. A synthetic dilution water (aerated distilled water+nutrient salts for the bacteria)[5] is now used instead of the former aerated tap water.

Many factors can influence this test, the most important being temperature variation, composition of dilution water, the dilution used, nitrification, the presence of toxic substances, the nature of the bacterial seed and the presence of anaerobic organisms.

It is important to keep the temperature of the B.O.D. incubator reasonably constant, i.e. to within ±0·5°C. It must be remembered that an increase (or decrease) of only 1°C means an increase (or decrease) in B.O.D. of nearly 5 per cent. The use of certain mineral salts in the dilution water (potassium and sodium phosphates, ammonium sulphate, magnesium sulphate, calcium chloride and ferric chloride) helps to maintain the proper salinity and buffer action and provides the necessary nutrients for the bacteria.

Since the B.O.D. may vary somewhat with the particular dilution used, with unknown samples it is advisable to put on several dilutions and to base the value on that showing about 40–50 per cent depletion of the dissolved oxygen.

Since the B.O.D. test does not distinguish between oxygen demand caused by carbonaceous oxidation and that due to nitrification (i.e. oxidation of ammonia to nitrite and nitrate), sewage effluents in a state of active nitrification (particularly activated-sludge plant effluents and some from alternating double filtration and recirculation plants) can show a misleadingly high B.O.D. and yet be quite satisfactory otherwise. For example, a sewage effluent had a 5-day B.O.D. of 31·4 mg/l, i.e. greater than the Royal Commission limit of 20 mg/l. But only 9·0 mg/l was due to actual carbonaceous matter and the remaining 22·4 mg/l was accounted for by the conversion of ammoniacal nitrogen to nitrite and nitrate by nitrifying organisms.

Nitrification during the B.O.D. test can in such cases be prevented by destroying the nitrifying organisms in the sample, e.g. by flash pasteurisation or by acidification to pH 2–3, followed by the usual B.O.D. procedure using dilution water seeded with settled domestic sewage.

A new and much simpler method of inhibiting nitrification consists in using dilution water containing 0·5 mg/l of allyl thiourea[11a,11b].

Acids, alkalis, toxic metals, free chlorine and many other bacterial poisons present in trade wastes have a marked effect on the B.O.D. test. For instance, as little as 0·01 mg/l of copper depresses the result by about 5 per cent. It is important, therefore, that the water used for dilution should be free from these toxic substances. It is equally evident that trade wastes containing appreciable amounts of bacterial poisons (e.g. gas liquor) will give misleadingly low B.O.D. results; in such cases the use of the 4 h permanganate test or the dichromate test is recommended to assess the amount of organic matter present.

Despite its limitations, the B.O.D. is probably the best single test of the strength of sewage and of many nitrogenous industrial wastes (e.g. dairy, slaughterhouse, food preparation wastes).

The 5-day B.O.D. of crude domestic sewage is normally about 3–5 times the 4 h P.V. Toxic chemical wastes, on the other hand, can give a ratio as low as 0·2. Industrial sewages containing inhibitory trade wastes may give a B.O.D./P.V. ratio intermediate between 0·2 and 3·0 depending on the proportion of trade wastes present. The B.O.D./P.V. ratio may, therefore, be regarded as giving a general indication of the relative ease with which substances present can be oxidised biochemically. Ratios equal to or greater than that for sewage indicate a waste water relatively easy to oxidise biochemically, ratios lower than 3·0 suggest possible difficulties in the biological treatment of the waste.

Typical B.O.D./P.V. ratios of some waste waters are given in *Table 6*.

Table 6. TYPICAL B.O.D./P.V. RATIOS FOR SOME WASTE WATERS

Waste water	5-day B.O.D. / 4 h P.V.	Biochemical oxidation
Birmingham domestic sewage	4·0	Easy
Birmingham industrial sewage	1·8	Rather difficult
Toxic chemical wastes	0·2	Very difficult
Dairy wastes	9·0–13·0	Very easy
Farm wastes	2·0–3·0	More difficult than sewage
Slaughterhouse wastes	7·0–11·0	Very easy

Plain sedimentation of sewage usually removes about 30–50 per cent of the B.O.D.; after aerobic biological purification, about 90–98 per cent should have been removed.

ORGANIC CARBON

The determination of organic carbon, which involves wet oxidation with sulphuric and chromic acids and volumetric estimation of the carbon dioxide evolved, is time-consuming, and the apparatus occupies much bench space. Hence the method is mainly of value for research purposes. The organic carbon figure, together with the total N figure, can be used to calulate the Ultimate Oxygen Demand of a sample (p. 30).

DICHROMATE VALUE OR 'C.O.D.'

Oxygen absorption from potassium dichromate, carried out in boiling 50 per cent H_2SO_4 for 2 h, using Ag_2SO_4 as catalyst, has now been put forward as a standard test, particularly for trade effluents. A troublesome interference in this test is chloride, but addition of mercuric sulphate complexes the chloride and so renders it inactive[5,5a].

Nature of Sewage and Chemical Analysis

The method can also be applied to sewages and tank effluents with oxygen demands of 50 mg/l or more, but not to sewage effluents. Sewage and most trade wastes are almost completely oxidised in the test. In general routine work, however, the method is unlikely to supplant the time-honoured and simpler permanganate test which is suitable for all types of wastes and effluents, strong and weak.

SUSPENDED SOLIDS

The determination of suspended solids is one of the most important of all the tests for sewages, tank effluents and final effluents. The removal of matter suspended in raw sewage is, indeed, one of the indications of the efficiency of settlement tanks: carefully operated, well designed settling tanks should remove about 50–90 per cent. But the efficiency of a settling tank can depend also upon the *nature* of the suspended matter. A limit of 30 mg/l of suspended solids was placed in the 8th Report of the Royal Commission on Sewage Disposal[1] on sewage effluents discharging to inland streams and receiving a dilution of at least 8 volumes of clean river water (p. 4).

SETTLEABLE SOLIDS

This very simple test can be carried out by settling litre samples of sewage, tank effluent and final effluent for 1 h in an Imhoff cone and estimating the volume of settled sludge in ml. Perhaps more accurate than the suspended solids test, it indicates the proportion of insoluble solid matter removed by sedimentation and thus serves as an index of the sludge-forming characteristics of the sewage. Typical average figures obtained for industrial sewage in an Imhoff cone are: crude sewage, 4·9; tank effluent, 0·35; final effluent*, 0·31 (all in ml/1).

TOTAL SOLIDS

The total solids, or residue on evaporation and drying, provides a rough index of the strength of a sewage, being in general higher with stronger sewages than with weaker. The reduction of total solids by treatment processes is an approximate measure of the efficiency of the treatment.

* Mixed effluent from activated sludge and percolating filter plants.

COMBINED NITROGEN

Nitrogen is present in crude sewage in organic form and as ammonia. During purification, organic N is broken down to ammonia, which is further oxidised to nitrites and nitrates. Evidently the relative proportions of the various forms of nitrogen indicate the degree of stabilisation of the organic matter. Instead of determining organic N it has been customary, in order to save time in routine work, to estimate albuminoid N, i.e. the fraction of the organic N which is easily decomposed when the sample (after being freed from ammonia by distillation) is distilled with alkaline potassium permanganate under standardised conditions.

The amount of organic (or albuminoid) N is an index of the amount of nitrogenous organic matter in the sewage, and of its 'strength'. Free ammonia is produced from organic nitrogenous substances by bacterial decomposition of sewage and may therefore tend to increase somewhat during sedimentation; too large an increase of ammoniacal N, however, usually indicates too long a detention period in the settling tanks.

Albuminoid N (or organic N) provides a good measure of the completeness with which purification of sewage has proceeded. In a satisfactory sewage effluent, the albuminoid N figure should not normally exceed 2 mg/l.

Nitrites are an intermediate stage in the bacterial oxidation of ammonia to nitrates

$$NH_3 + 3O = HNO_2 + H_2O$$
$$HNO_2 + O = HNO_3$$

Well purified, percolating filter effluents usually contain only small amounts of nitrite N (1 mg/l or less), and one of the first signs of deterioration in a filter (due to overloading, or to 'ponding') is generally an increase in nitrite N and a fall in nitrate N.

Nitrites and nitrates are not usually found in raw sewage, except in wet weather and in the presence of certain trade wastes. Since nitrates are the final oxidation products of ammonia, they are indications of effluent stability and of the efficiency of biological oxidation. In well operated conventional filters, 50 per cent or more of the ammoniacal N may be oxidised to nitrate, and except with weak or wet-weather sewage, this means generally a nitrate N content of 10 mg/l or more. The proportion of ammonia oxidised is less in filters operated by recirculation or by alternating double filtration, especially during the winter when low temperatures tend to inhibit the activities of nitrifying organisms. Activated sludge plants, however, are generally designed to give only carbonaceous oxidation of sewage, since a rather long aeration period, and consequently a high power consumption, would be needed for extensive nitrification of the sewage. Activated sludge plant effluents, therefore, contain much higher concentrations of ammoniacal nitrogen than percolating filter effluents.

CHLORIDE

The amount of chloride over and above that of the water supply of the area can be used as an index of the strength of a sewage. Certain trade wastes (e.g. from meat-packing and water-softening plants) may, however, considerably increase the chloride content of sewage. Chloride is unaffected by sewage treatment processes; samples taken at various stages of the purification of sewage (e.g. raw sewage, tank effluent and final effluent) should, therefore, have about the same chloride content if truly comparable. Dilution of sewage during wet weather causes a fall in the chloride figure; comparison with the normal dry-weather chloride content may be used to give a rough idea of the degree of dilution.

pH, ACIDITY AND ALKALINITY

The intensity of the acidity or alkalinity of an aqueous solution depends upon the hydrogen ion concentration. Owing to the very wide range of H ion concentrations and to the extremely low values generally encountered, Sörensen first suggested that it would be more convenient in practice to use the more compact logarithmic scale of pH values based on the relationship

$$\mathrm{pH} = -\log_{10} C_H = \frac{1}{\log_{10} C_H}$$

where C_H = hydrogen ion concentration in g/l. Thus, if the H ion concentration is 10^{-5} ($1/10^5$), the pH value is 5. It is important to note that the pH scale, which can extend from $-0\cdot3$ (6N HCl) to $14\cdot5$ (7N KOH), is a logarithmic, not an arithmetical, scale, Thus, an increase in pH of 1 unit means a tenfold decrease in hydrogen ion concentration (i.e. in the intensity of acidity), and so a solution of pH value 6 has 10 times the concentration of H ions of a solution of pH 7. Pure water, which is neutral, has a pH value of 7, and values above or below 7 indicate alkalinity and acidity, respectively. Normal domestic sewage is usually very slightly alkaline (pH $7\cdot2$–$7\cdot4$) but, as it is well buffered, the presence of industrial wastes does not usually have a marked effect on pH unless these are strongly acid or alkaline and present in considerable quantity. If the pH of the sewage is less than 5 or greater than 10, there may be considerable interference with aerobic biological processes. The determination of pH is usually carried out using organic indicators, but for more accurare results, a small portable pH meter is available.

The alkalinity to methyl orange falls during biological treatment of sewage on account of production of CO_2 by oxidation of carbon compounds and of nitric acid by oxidation of ammonia.

STABILITY

Stable effluents containing dissolved oxygen and nitrate (a source of combined oxygen) remain aerobic for a considerable period of time.

The stability of sewage effluent (or its 'keeping quality') is its ability to maintain itself in an oxidised state when kept out of contact with oxygen or air in an incubator at a specified temperature. Many forms of incubator tests have been described. One of the simplest involves addition of methylene blue (1·33 mg/l) which can be obtained in tablet form; the sample is considered to have failed the test if the dye is decolorised within 5 days at 20°C. After this period unstable effluents will generally contain H_2S. Percolating filter effluents containing nitrate almost invariably pass a stability test. Since a normal activated-sludge plant effluent contains little or no nitrate, it may fail the stability test, unless it contains much dissolved oxygen and the B.O.D. and suspended solids are well within the Royal Commission limits of 20 mg/l and 30 mg/l, respectively.

SIMPLE METHODS FOR TESTING SEWAGE EFFLUENTS

The Ministry of Technology has issued a leaflet (*Notes on Water Pollution No, 44*) giving a number of such simple methods suitable for a small sewage works. Tests include suspended solids, settleable solids, pH, B.O.D., C.O.D., permanganate value, stability and ammonia.

STRENGTH OF SEWAGE

The size of a Sewage Treatment Works depends not only upon the volume of sewage to be treated but also upon its 'treatability' (the ease with which is can be purified) and 'strength', i.e. its content of organic and other oxidisable matter. Sewages from different towns show considerable variations in the content of organic and other oxidisable matter, and so differ in strength. This is generally measured by the amount of oxygen required to oxidise the organic and inorganic constituents of a given volume of sewage. The treatability of the sewage is not necessarily related to its strength but depends upon the nature of the substances present and their toxicity to micro-organisms. Thus, many synthetic organic substances are more difficult to oxidise biologically than naturally occurring organic materials. Again, metallic trade wastes although not affecting the strength of sewage because they have no oxygen demand, nevertheless greatly increase the difficulties of treatment because of their toxicity.

The following are the most important factors which influence the strength of a sewage.

1. *Type of sewerage system*—In a combined system taking all storm water, the sewage will be relatively weak during heavy rain but strong during very dry weather.
2. *Infiltration water*—Water leaking into sewers through defective sewer fabric and joints will, in general, tend to dilute the sewage.
3. *Daily water consumption*—The average water consumption in towns in England and Wales is around 180 l (40 gal) per head per day. This average is rather higher in Scotland and very much higher in the U.S.A. American water consumption may be as much as 450–900 l (100–200 gal) per head per day, and so American sewages are usually weaker than English ones.
4. *Food and other habits of the population*—The food and other habits of the population and the nature of the sanitary accommodation can have a marked effect upon the strength of sewage. For instance, sewage from military camps where the diet contains a high proportion of meat, fat and other foods of high calorific value, and where modern methods of vegetable preparation are used, tends to be much more organic in character and hence stronger than sewage from purely residential districts; the strength is also increased by the use of the separate system. Again, areas in which there is a preponderance of waste water closets tend to have a sewage stronger than those areas served entirely by W.C.s, the reason being that in the former case the water is virtually used twice.
5. *Nature and quantity of trade wastes*—Trade wastes have a big influence on the strength of a sewage and on its amenability to treatment.

Sewage strength can be assessed by many of the tests currently used for measuring oxygen demand, such as B.O.D., permanganate value and dichromate value, as well as by determinations of organic carbon, albuminoid or organic nitrogen, ammoniacal nitrogen, chloride and total solids.

A method of assessing sewage strength much used in Britain is that first suggested by McGowan[1] in the 5th Report (Appendix IV) of the Royal Commission and stricly applicable only to domestic sewage:

Strength of sewage = (ammoniacal nitrogen + organic nitrogen) × 4·5

+ (N/8 permanganate value × 6·5)*

all the figures being expressed in parts of 100 000. The factor 4·5 represents the quantity of oxygen required to oxidise the nitrogen to nitrate. The factor 6·5 is really applicable only to domestic sewage since it is based on the proportion of sewage organic matter oxidised by the permanganate under the conditions of the test. When dealing with gas liquor, for example, a different factor would have to be used to determine the McGowan strength (1·4 instead of 6·5).

The permanganate value itself can also be used as a rough measure of sewage strength. The N/8 permanganate value expressed in mg/l is very approximately the same as the McGowan strength figure expressed in parts per 100 000.

The 5-day B.O.D. is often used to indicate approximately the strength of a sewage and is specially useful for domestic sewage and for nitrogenous organic trade wastes. Toxic trade wastes and sewages containing large amounts of inhibitory trade wastes, however, yield misleading results. The 5-day B.O.D. can perhaps better be regarded as an indication of the ease with which biochemical oxidation takes place. A high B.O.D. : 4 h permanganate value ratio indicates a sewage easily oxidisable biochemically whilst a low ratio denotes a sewage more difficult to oxidise. A better measure of sewage strength than the 5-day B.O.D. is the long-term B.O.D. or Ultimate Oxygen Demand (U.O.D.)[12], i.e. the B.O.D. as determined after 2–3 months' incubation when virtually complete oxidation will have occurred. This is a tedious estimation even if done by respirometer methods. It can, however, be calculated approximately by the formula

$$\text{U.O.D.} = 2\cdot67\ C + 4\cdot57\ N$$

---

* See footnote on p. 21.
If the results are in mg/l, the formula becomes 0·45 (ammoniacal N + organic N) + 0·65 N/8 P.V.

Nature of Sewage and Chemical Analysis

where C = organic carbon in mg/l, N = sum of ammoniacal and organic nitrogen (both in mg/l).

The determination of dichromate value or C.O.D. can also be used to give a good approximation to the U.O.D.

The approximate relationship between McGowan strength, 5-day B.O.D. and permanganate value is shown in *Table 7*.

Table 7. APPROXIMATE RELATION BETWEEN McGOWAN STRENGTH, 5-DAY B.O.D. AND N/8 PERMANGANATE VALUE FOR DOMESTIC SEWAGE

Sewage	McGowan strength parts per 100 000	5-day B.O.D. (20° C) mg/l	N/8 permanganate value mg/l
Weak	60	210	50
Average	100	350	100
Strong	170	600	150

These figures can be used for the approximate conversion of a trade waste to its equivalent of sewage. For example, if 45 000 l (10 000 gal) of a trade waste having a 4 h N/8 permanganate value of 2 000 mg/l is admitted to the sewers, the waste is 2 000/100 = 20 times as strong as an average domestic sewage and will be roughly equivalent to an additional 45 000 × 20 = 900 000 l (200 000 gal) of average domestic sewage.

In the U.S.A., the strength of trade wastes admitted to the sewers is often expressed in terms of the equivalent population. A dry-weather American sewage has a 5-day B.O.D. averaging 0·077 kg (0·17 lb) per head of population per day. Hence, the population equivalent of an industrial sewage or trade waste is given by the expression

Population equivalent (U.S.A.)

$$= \frac{\text{lb. of 5-day B.O.D. per day from industrial sewage or trade waste}}{\text{lb. of 5-day B.O.D. from domestic sewage per day per head of population}}$$

$$= \frac{\text{5-day B.O.D. (mg/l)} \times \text{flow (in } 10^6 \text{ gal/day)} \times 8\cdot34}{0\cdot17}$$

since a U.S.A. gallon of water weighs 8·34 lb.

The corresponding British 5-day *per capita* B.O.D. was for many years thought to be 0·054 kg (0·12 lb), but the figure is now probably nearer 0·073 kg (0·16 lb) for crude sewage (or for settled sewage, 0·045 kg or 0·10 lb).

The British population equivalent is, therefore, given by the expression

Population equivalent (British)

$$= \frac{\text{5-day B.O.D. (mg/l)} \times \text{flow (m}^3)}{0\cdot073 \times 10^3}$$

TYPICAL ANALYSES OF SEWAGES AND SLUDGES

Table 8 gives typical analyses of weak, average and strong domestic sewages.

Table 8. TYPICAL ANALYSES OF WEAK, AVERAGE AND STRONG DOMESTIC SEWAGES

Determination mg/l	Sewage weak	Sewage average	Sewage strong
4 h N/8 permanganate value	50	100	150
5-day B.O.D., crude	210	350	600
5-day B.O.D. settled	150	250	400
Suspended solids	200	350	500
Ammoniacal nitrogen	25	40	90
Albuminoid nitrogen	5	10	15
Chloride	60	100	140

The values given in this table may, of course, be markedly affected by the presence of industrial wastes.

Table 9 shows the analytical results at the various stages of purification of a strong industrial sewage containing gas liquor when treated in settling tanks, percolating filters and humus tanks.

Some important features of these results are:

(a) The sewage is very strong, judged by the 4 h permanganate value, but only about average on the B.O.D. test; this is typical of sewages containing gas liquor and certain other toxic wastes.
(b) The 4 h permanganate value of the crude sewage and of the tank effluent is about twice the 3 min value. The normal 4 h : 3 min ratio for domestic sewage is 3, but the presence of

Table 9. STAGES OF PURIFICATION OF INDUSTRIAL SEWAGE CONTAINING GAS LIQUOR (BIOLOGICAL FILTRATION)

Where taken	pH	M.B. stability test 5 days F = fails P = passes	Transparency seen through mm shaken	Transparency seen through mm settled	N/8 permanganate value 3 min	N/8 permanganate value 4 h	Suspended matter min	Suspended matter vol.	Suspended matter total	Nitrite N	Nitrate N	B.O.D. 5 days 20°C	Phenols (expressed as cresols)
Crude sewage	7·5	F. 15 min (H₂S)	34	66	99·0	208·0	37	120	157	nil	nil	347·0	17·5
Settling tank effluent	7·0	F. 15 min (H₂S)	55	76	38·0	76·0	13	22	35	nil	nil	173·0	8·75
Percolating filter effluent	7·0	P	110	270	12·4	34·4	20	60	80	nil	6·4	37·1	1·25
Final effluent (humus tanks)	7·0	P	205	285	8·4	22·4	4	7	11	nil	5·6	12·4	1·25

Table 10. STAGES OF PURIFICATION OF AN INDUSTRIAL SEWAGE (ACTIVATED SLUDGE-SHEFFIELD BIO-AERATION PROCESS)

Where taken	N/8 Permanganate value 4 h	Chloride as Cl	Suspended matter	Nitrite N	Nitrate N	B.O.D. 5 days, 20°C	Nitrogen Ammonia	Nitrogen Albuminoid
Crude sewage	121·3	341	226	nil	nil	262·0	22·6	6·97
Tank effluent	93·8	333	86	nil	nil	200·4	23·8	5·13
Final effluent	28·0	309	22	nil	nil	17·4	22·8	1·92

easily oxidisable substances (phenols and other constituents of gas liquor) tends to lower this ratio.
(c) Nitrate is absent from the crude sewage and tank effluent but present in the final effluent.
(d) Good removal of suspended solids occurs in the settling tanks, amounting to 78 per cent; the overall removal of suspended solids, from raw sewage to final effluent, is 93 per cent.
(e) The effluent from the percolating filters contains much more suspended matter than that from the settling tanks, owing to the presence of humus sludge flushed from the filters. This is subsequently removed by settlement in humus tanks, giving a final effluent low in suspended solids.
(f) The reduction in B.O.D. is 50 per cent in the settling tanks, the overall reduction, from crude sewage to final effluent, is 96 per cent, which is very good.
(g) The overall reduction of phenols in the plant, of 93 per cent, is also good.
(h) The final effluent is stable and passes the Royal Commission standards for B.O.D. and suspended solids. It is to be noted, however, that its 4 h permanganate value is much greater than the B.O.D. and above 20 mg/l. This is a common feature of effluents derived from gas liquor and certain other trade wastes, caused by the presence in the effluent of substances resistant to biochemical oxidation (e.g. higher tar acids from gas liquor) but relatively easily oxidised by permanganate.

Table 10 gives analyses obtained in an activated-sludge plant treating an industrial sewage. It will be seen that there are good reductions in B.O.D. and suspended solids, and that the final effluent conforms to Royal Commission standards but has an abnormally high 4 h permanganate value, owing to the presence of residual trade waste constituents resistant to biochemical oxidation but easily oxidised by permanganate. There is no nitrate in the final effluent—a feature of most activated-sludge plant effluents—and, although the albumonoid nitrogen is considerably reduced, the ammoniacal nitrogen remains unchanged. The chloride figures show reasonable constancy, as chloride is unaffected by sewage treatment.

A fairly complete analysis of an average strength domestic sewage is shown in *Table 11*. The corresponding figures to be expected of the final effluent from this sewage treated by an aerobic purification process are also given.

Table 12 gives typical analyses of the commoner types of sewage sludge. These are of interest in view of the use of sewage sludges as fertilisers. The pH value of raw primary sludge is usually within the range 6·5–7·0 but soon falls on storage; well digested sludges usually have a pH value between 7·0 and 7·8.

In addition to the chemical constituents listed in *Table 12*, sewage sludge also contains considerable amounts of mineral matter, especially silicon, iron, aluminium, calcium, magnesium and sulphur as well as traces of many minor elements (e.g. manganese, zinc, copper and boron) and of organic growth-promoting substances. Many of these substances are essential for plant growth.

Table 11. TYPICAL ANALYSES OF AN AVERAGE STRENGTH DOMESTIC SEWAGE AND FINAL EFFLUENT (AEROBIC PURIFICATION PROCESS)

		Raw sewage	Final effluent
3 min P.V.	mg/l	34	5
4 h P.V.	mg/l	100	15
Dichromate Value	mg/l	500	40
5-day B.O.D.	mg/l	350	20 or less
Suspended Solids	mg/l	350	30 or less
Total Solids	mg/l	900	–
Ammoniacal N	mg/l	35	20 or less*
Organic N	mg/l	20	4
Albuminoid N	mg/l	10	2
Chloride	mg/l of Cl'	100	100
Phosphorus	mg/l of P	15	7
Potassium	mg/l of K	12	10
Alkalinity to M.O.	mg/l of $CaCO_3$	400	250
pH range		7–8	6–8
Anionic Synthetic Detergent mg/l of Manoxol O.T.		15	6 (from 'hard' detergent) 2–3 (from 'soft' detergent)
Presumptive coliform count per 100 ml		350 000	3 000†
Total bacterial colonies per ml (3 days, 22°C)		5 000 000	100 000

* In a normal percolating filter, where much nitrate N is produced, ammoniacal N should be quite low.
† It should be noticed that this figure is high compared with that of an acceptable drinking water, where the numbers per 100 ml are in accordance with the following Ministry of Health classification[13]:

Excellent	0
Satisfactory	1–3
Suspicious	4–10
Unsatisfactory	more than 10

The gas evolved during sludge digestion generally contains about 70 per cent methane and 30 per cent carbon dioxide by volume, with traces of nitrogen, hydrogen and hydrogen sulphide. Its fuel value is about 22 500–26 000 kJ/m^3 (600–700 Btu/ft^3), i.e. higher than ordinary coal gas 17 000–22 000 kJ/m^3 (450–600 Btu/ft^3). Sludge

Table 12. TYPICAL ANALYSES OF SEWAGE SLUDGES

Type of sludge	Moisture per cent	\multicolumn{5}{c}{Analysis on dry basis per cent}
		N	K_2O	P_2O_5	Grease (light petroleum extract)	Organic matter
Sedimentation tank (primary)	90–95	2·5–4·5	—	1–3	20–35	60–80
Humus	92–98	5–7	—	2–3·5	5–10	55–70
Activated	98–99·5	5–7	0·1–0·9	2–3·8	5–10	65–75
Digested	88–95	1·5–3·0	0·1–0·4	1–2	3–8	45–60

digester gas is used to generate power at many sewage works (especially the larger activated-sludge plants). Some typical analyses are shown in *Table 13*.

Sludge gas forms explosive mixtures with air (1 part of sludge gas with about 7–18 parts of air). Smoking is, therefore, forbidden in and near sludge digestion plants and, since a spark from a tool can cause an explosion, the use of brass and bronze tools is advised.

An interesting and comprehensive paper dealing with the properties, detection and methods of analysis of sludge gas is given by Burgess and Wood[14]

Table 13. COMPOSITION OF SLUDGE DIGESTER GAS

	I	II	III	IV	V	VI
CH_4	64·3	61·2	72·9	67	68·4	68·6
CO_2 per cent	32·6	32·3	24·6	30	31·0	29·9
N by	1·7	2·3	1·6	3	0·6	1·2
O volume	—	0·6	0·6	—	—	0·17
H_2S	—	—	0·3	—	—	0·13
H	1·4	3·5	—	—	—	—
kJ/m^3	21 910	23 800	27 000	23 290	25 450	22 920
Btu/ft^3	588	639	725	625	683	615

REFERENCES

1. Royal Commission on Sewage Disposal, Report No. 8. London; H.M.S.O.
2. INKSTER, J. E., Oxygen balance in polluted water. *J. Proc. Inst. Sew. Purif.* (1943) 123
3. *Ministry of Housing and Local Government, Methods of chemical analysis as applied to sewage and sewage effluents.* 2nd Ed., 1956. London; H.M.S.O. (to be revised)
4. *Joint Committee, Association of British Chemical Manufacturers and Society for Analytical Chemistry, Recommended methods for the analysis of trade effluents,* 1958. London; Society for Analytical Chemistry
5. *Standard methods for the examination of water and wastewater,* 12th Ed., 1965. New York; American Public Health Association
5a. JENKINS, S. H., *et al* Some analytical methods used in the examination of sewage and trade wastes. *J. Proc. Inst. Sew. Purif.* 6 (1965) 533
6. THEROUX, F. R., ELDRIDGE, E. F. and MALLMANN, W. L., *Laboratory manual for chemical and bacterial analysis of water and sewage.* 3rd Ed., 1943. London and New York; McGraw-Hill
7. LESTER, W. F. and RAYBOULD, R. D., Some notes on sewage analysis. Part II. The analysis of sewage sludge, *J. Proc. Inst. Sew. Purif.* 4 (1950) 398
8. WISHART, J. M., JEPSON, C. and KLEIN, L., Dewatering of sewage sludge by coagulation and vacuum filtration. Part I. Laboratory experiments (Buchner funnel test), *J. Proc. Inst. Sew. Purif.* 1 (1946) 110
9. COACKLEY, P., Research on sewage sludge carried out in the Civil Engineering Dept. of University College, London (Specific Resistance Test) J. *Proc. Inst. Sew. Purif.* 1 (1955) 59
10. JONES, B. R. S. and JENKINS, S. H., Filtration of sewage sludge from Yardley Works (Specific Resistance and Filter leaf tests), *J. Proc. Inst. Sew. Purif.* 4 (1955) 279
11. ARDERN, E. and LOCKETT, W. T., Laboratory tests for ascertaining the condition of activated sludge, *J. Proc. Inst. Sew. Purif.* 1 (1936) 212
11a. WOOD, L. B. and MORRIS, H., Modifications to the B.O.D. test (use of allyl thiourea). *J. Proc. Inst. Sew. Purif.* 4 (1966) 350
11b. MONTGOMERY, H. A. C. and BORNE, B. J., The inhibition of nitrification in the B.O.D. test (use of allyl thiourea). *J. Proc. Inst. Sew Purif.* 4 (1966) 357.
12. GAMESON, A. L. H. and WHEATLAND, A. B., Ultimate oxygen demand and course of oxidation of sewage effluents, *J. Proc. Inst. Sew. Purif.* 2 (1958) 106
13. *Ministry of Health and Ministry of Housing and Local Government. The Bacteriological examination of water supplies.* Report No. 71, 1956. London; H.M.S.O.
14. BURGESS, S. G. and WOOD, L. B., The properties and detection of sludge gas. *J. Proc. Inst. Sew. Purif.* 1 (1964) 24

Chapter 3

SEWERAGE SYSTEMS, STORM SEWAGE OVERFLOWS AND STORM SEWAGE TREATMENT

SEWERAGE SYSTEMS

Before sewage can receive treatment it has to be conveyed to the sewage works by means of a sewerage system. It is first discharged via drains into local sewers which are connected to district sewers, and these in turn discharge into trunk sewers which convey the sewage to the treatment works. In dry weather the liquor in the sewers consists of one or more of the following constituents, in proportions varying with local conditions and the state of the sewers:

(a) *Domestic Sewage*: waste water from W.C.s, baths, sinks, wash-basins and wash-houses
(b) *Trade Waste*: the waste waters from industrial processes
(c) *Infiltration Water*: i.e. ground water entering the sewers owing to pervious sewer fabrics and faulty joints, and sometimes to land drains having been connected to the sewers.

The flow in sewers varies considerably throughout the day, even in dry weather (see *Figure 3*), owing to a number of factors which vary from district to district. The flow of domestic sewage is generally several times greater in volume at midday than in the early hours of the morning; the difference is less in districts where many industries are engaged in shift work throughout the 24 hours of the day. Discharges of trade waste considerably affect the flow in sewers; in some cases the liquor is discharged at a uniform rate throughout the 24 hours, resulting in a more even flow in the sewers, but where the liquors are discharged in flushes, or only during daytime, variation of flow in the sewers increases. The flow of infiltration water is generally constant throughout the day but can be affected by weather conditions.

Sewerage systems can generally be classified in one of the following categories, based on the arrangements made for dealing with surface water in wet weather.

(a) *Combined System*—the foul sewage and surface water from roads, streets, yards and roofs discharge into a common sewer for conveyance to the sewage works for treatment
(b) *Separate System*—the sewage discharges into a foul sewer which conveys it to the sewage works for treatment; all the surface water runs into a separate surface-water sewer and is discharged into the nearest suitable watercourse
(c) *Partially Separate System*—most of the surface water (generally water from roads, paths and roofs of buildings) runs into a surface water sewer and is discharged into the nearest watercourse; the foul sewage plus some surface water (generally water from

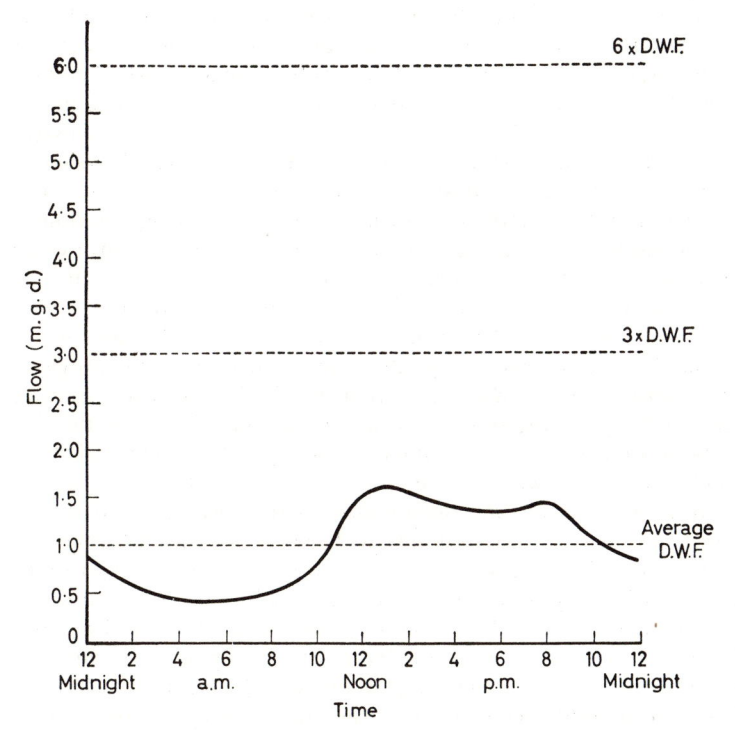

Figure 3. Typical flow chart for domestic sewage in dry weather (D.W.F. 1×10^6 gal/day)

back roofs of buildings) discharges into the foul sewer which conveys it to the sewage works for treatment.

In towns which have increased in size over a period of years the sewerage system is quite often a combination of all three types.

In combined sewerage systems, a tremendous volume of water enters the common sewer in times of very wet weather, and it would be impracticable to construct sewers large enough to convey all this water to the sewage works. Consequently, storm water overflows are constructed on the sewers to allow all sewage over a certain flow to be discharged to the nearest suitable watercourse. These overflows are often sited on the district sewers just above the junction point with the trunk sewer, but the selection of a site for an overflow may have to be determined by the proximity of a suitable watercourse. Partially separate systems sometimes require the provision of overflows on the foul sewers, depending upon the proportion of surface water which is allowed to discharge into the foul sewer. Although the flow in the foul sewers of separate sewerage systems tends to increase somewhat in wet weather, there should be no need for the provision of overflows, providing the foul sewers are made large enough to carry 4 to 6 times the normal dry-weather flow.

Each of the above types of sewerage systems has its advantages and disadvantages. The comparative costs are to a great extent dependent upon geographical factors. In a combined system a larger sewer has to be constructed, but with a separate system two are required, the length of the surface water sewer being determined by the distance from the nearest receiving watercourse. Storm water overflows have to be constructed on combined sewers resulting in polluting storm water being discharged into watercourses. Such overflows are obviated by using a separate system. In cases where the land formation necessitates the pumping of sewage, the provision of a separate system can result in a reduction in the capital and running costs of the pumping stations, owing to the smaller volume that has to be dealt with. A disadvantage of the separate system is that the discharge into a watercourse from a surface water sewer in wet weather is anything but clean water, as the surfaces of roads and streets are heavily contaminated with oil, grease, solid matter, etc. This may well be aggravated if the idea of washing and brushing particularly dirty streets with a detergent solution is developed and extended, for this could result in streams being polluted by discharges from surface water sewers in dry weather. Separate systems also have the disadvantage that there is always the danger of foul sewage being accidentally connected to the surface water sewer, and in industrial areas spillage of industrial liquids in works yards could result in polluting discharges passing into a watercourse via a surface water sewer. Discharge of sewage from foul sewers into surface water sewers can also take place on separate systems incorporating dual-purpose manholes. It is evident that careful consideration has to be given to many factors when deciding upon the type of sewage system to be adopted.

STORM WATER OVERFLOWS ON SEWERS

As stated previously, storm water overflows have to be constructed on those sewers receiving surface water in addition to sewage. Decision on the flow at which these overflows shall come into operation is very important, as serious river pollution can be caused if they are set at too low a flow. When a storm commences, the flow in the sewers starts to increase, and it has been proved that during the first few hours of a storm the suspended solids content of the sewage is much above the normal dry-weather level, owing to the redispersion of matter which has settled out in the sewer. The flow in the streams also increases during wet weather, giving a greater dilution to the discharges from the storm water overflows, but scour, etc. may cause a deterioration in the quality of the stream water. Where there is a combined sewerage system in a heavily built-up district with large areas of impervious road surfaces, the flow in the sewers tends to rise at a more rapid rate than that in the watercourses, with the result that the overflows will commence to operate before there has been a comparable rise in the stream flows.

Over the years it had become the prevalent practice to construct overflows on sewers so that they come into operation when the flow exceeds 6 times the average daily dry-weather flow; the figure of 6 × D.W.F. was the recommendation of the then Local Government Board in 1909. This differs somewhat from the recommendations of the Royal Commission on Sewage Disposal[1], which (on page 7 of its Final Report) said that storm overflows on branch sewers should be set so as not to come into operation until the flow in the branch sewers was several times the maximum normal dry-weather flow in the sewer.

The average daily dry-weather flow at any point (avg. D.W.F.) can be defined simply as the average flow of sewage, expressed as gal/day, which passes that point during a period of 24 h under dry weather conditions.

There are varying opinions as to the best method of calculating the average D.W.F. especially with regard to what constitutes 'dry weather' conditions. For example, one local authority calculates the average D.W.F. at its sewage works by taking an average over the year of daily flows of sewage on those working days when the daily rainfall does not exceed 1 mm (0·04 in), after a period of 7 consecutive days with less than 1 mm (0·04 in) on any one day. Another authority determines the average daily flow of sewage over a period of at least one week following 14 consecutive dry days in summer. Other methods of arriving at the average D.W.F. have also been used.

The great fluctuations in sewage flows have been mentioned previously, and it can be seen that if an overflow is set at 6 × D.W.F. and a storm occurs when the sewage flow is at its peak, the sewage will have been diluted less than six times when the overflow starts to operate. This trouble is accentuated by the fact that at peak flows the sewage is quite often at peak strength, especially in those cases where trade wastes are discharged into the sewers. This is overcome by some local authorities by assessing the D.W.F. to be twice the flow between 8 a.m. and 8 p.m.

In some towns the practice has been introduced of setting the overflows at 6 × D.W.F. for the domestic sewage content, but using smaller multiples for the trade waste and infiltration water (e.g. 1·1 × trade waste and 1·0 × infiltration water). This is unsatisfactory, since it causes the overflows to operate at a flow less than 6 × D.W.F. of all the sewer contents. In addition it is illogical to allow a smaller dilution for trade waste, as this is generally much stronger than domestic sewage and quite often has the effect of increasing the peak flows in the sewer. Infiltration water can, at times, be anything but innocuous, an example being an area where it contains so much iron that the sewage has to be treated with chemicals before it receives biological treatment. Moreover, it is very difficult to determine accurately the proportion of infiltration water in the sewer contents.

There have been instances, especially in recent years, where overflows have been set at a flow calculated on a 'gallons per head of population per day' basis. Where this practice is adopted, a survey is made of the conditions in the drainage area, and a figure is arrived at for the estimated daily amount of sewage per head of population, including trade wastes in addition to domestic sewage. This figure is then multiplied by an agreed factor (usually 6), an addition is made for infiltration water, and overflows are set at a flow equal to the final calculated figure multiplied by the number of persons living in the drainage area above the overflow. Where a generously calculated figure results, overflows set on this basis can be quite satisfactory. If a uniform figure is used throughout a large drainage area, this practice is open to criticism, because in the industrial parts of the area a situation can arise where owing to large discharges of trade wastes and a small resident population, flow in the district sewer will be disproportionate, with the result that the overflow will operate prematurely. This trouble can possibly be overcome by using a 'population equivalent' figure (see p. 32).

It can be seen that the setting of storm water overflows is a complex problem, and taking all things into consideration it does appear that it might be preferable to abandon the idea of uniform settings for all overflows and to treat each overflow individually on the merits of local conditions. In fact, storm overflow settings higher than the customary $6 \times$ D.W.F. (e.g. $8 \times$ D.W.F., $10 \times$ D.W.F. or even $14 \times$ D.W.F.) have been sanctioned by the Ministry in certain sewerage schemes. Such higher settings may be justified, for example, on fishing streams where dilution by stream water is inadequate, or in other special circumstances.

To arrive at the setting of any storm water overflow on a sewer, due regard should be paid to

(a) The flow, nature, and composition of the sewage
(b) The capacity of the available sewers
(c) The condition, volume and velocity of the stream into which storm water will discharge, and the rate of run-off of the stream's catchment area
(d) The use to which the receiving stream will be put
(e) The impermeability of the area draining to the sewer.

There are many designs of overflows, some satisfactory and some bad. Experience has shown that the old-fashioned 'leap weir' is unsatisfactory and can give rise to a great deal of trouble. The side-weir type of overflow (either single or double-sided) (see *Figure 4*) has been widely used, but it has its disadvantages. It is rarely possible,

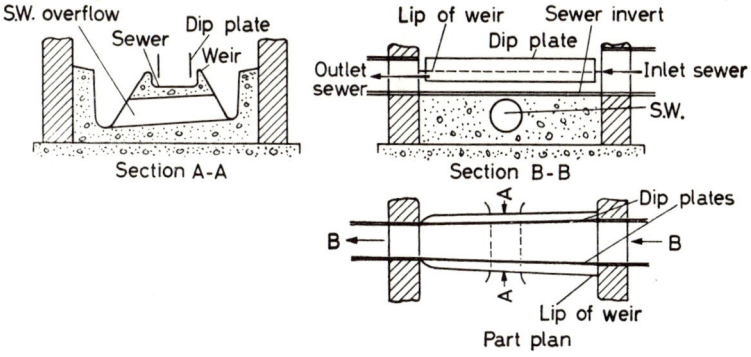

Figure 4. Storm water overflow

due to space restrictions, to provide a long length of weir, which makes it difficult to effect efficient separation at the required rate of flow. In addition there is a tendency for the overflowing storm sewage to contain appreciable amounts of solid matter, but the use of dip-plates can reduce the amount of gross solids to some extent. Special screens are on the market which can be fitted to the overflows and through which passes the overflowing storm sewage before discharge to the stream. These are automatically raked and the retained solids are returned to the flow of sewage passing forward down the sewer.

The 'stilling-chamber' type of overflow has become popular in recent years and some overflows of this type have been constructed incorporating a siphon for the discharge of storm sewage to the stream. The 'stilling chamber' together with the sewer upstream from the overflow act as a storage tank, with the result that the first foul flush has a better chance of passing forward to the treatment works and, in addition, the flow in the receiving stream has a chance to increase before the overflow starts to operate. This type of overflow utilises a throttle-pipe and is not very satisfactory on small sewers, as the throttle-pipe has then, of necessity, to have a small diameter, which is susceptible to blockage. It is desirable that the diameter of the throttle-pipe should never be less than 150 mm (6 in).

STORM WATER TREATMENT

All sewage passing the storm water overflows on the sewers eventually arrives at the sewage treatment works, and in wet weather obviously amounts to several times the normal D.W.F. It has been accepted that in most cases it is not practicable to give full treatment to all this sewage. The generally adopted procedure is to provide a separation weir on the works which allows a certain volume to pass forward for full treatment; the rest of the flow passes over the weir and is given partial treatment.

The usual practice has been to set the separation weir at a level which allows a flow of up to $3 \times$ D.W.F. to receive full treatment. This is in accordance with the Royal Commission's recommendation[1] that in wet weather a flow of about three times the normal dry-weather rate could be passed through ordinary sedimentation tanks without serious disadvantage.

There have been instances where multiples lower than three have been used for the trade waste and infiltration water in the sewage when arriving at a setting for the separation weir. This is undesirable, on grounds similar to those used against this type of practice for the setting of storm water overflows on sewers.

The usual position for the separation weir is just after the screens and detritus tanks, as this prevents gross floating solids and grit from entering the partial treatment system.

Separation weirs are in most cases of the side-weir type. The longer a weir is, the more accurately it can be set at a certain flow, but it is not always possible to construct a weir as long as one would like owing to the increase in cost and lack of available space. To overcome this difficulty a type of weir was developed which consists of a number of short channels blocked at one end, branching at right angles from the sewage carrier channel. The sides of these channels act as overflow weirs, thus giving the effect of a long weir in a comparatively small space.

In the early days of sewage treatment the storm water overflowing at the separation weir was treated on stand-by filters known as storm filters, but these were usually inefficient, and the Royal Commission[1] recommended the use of sedimentation tanks in their place. The use of sedimentation tanks for the treatment of storm water is now general practice, and it is usual to provide two or more tanks of a total capacity of $\frac{1}{4}$ D.W.F. This is based on the assumption that flows between 3 and 6 times D.W.F. will have to be treated, thus giving 2 h retention time in the tanks at maximum flow. Where the storm water overflows on the sewers are set at a flow greater than $6 \times$ D.W.F. and the separation weir at the works at 3 D.W.F., the capacity of the storm water tanks will have to be greater than $\frac{1}{4}$ D.W.F. to ensure 2 h retention time at maximum flow.

Storm water tanks are usually of the rectangular lateral flow type, although in some modern works circular sedimentation tanks have been used. As sedimentation tanks will be dealt with in Chapter 5 it is not proposed to discuss the design of storm tanks in detail here. In many cases the inlets to the storm tanks are set at slightly different levels but the outlet cills at a uniform level, with the result that the tanks fill up one after the other, but all commence to overflow at the same time if the storm flow continues.

The purpose of storm water tanks is not merely to give settlement to the storm sewage before it discharges into a watercourse; they are also retention tanks in which the storm sewage can be stored until it can be returned to the works inlet for full treatment when the flow of sewage has returned to normal. These tanks are therefore provided with means whereby the supernatant water can be returned for full treatment, which should be done as soon as possible after the flow has returned to normal. The sludge must also be removed from the tanks, so that they will be empty in readiness for the next storm. In view of this it is not desirable to make provision for the removal of sludge from storm tanks under hydrostatic head because in the event of a short storm the tanks may be only partially filled and it would then not be possible to remove sludge from the tanks.

The foulness of the first flush of sewage when a storm commences has been mentioned previously, and some sewage works' designers consider it advisable to provide an extra storm water tank with no outlet, which will fill before the storm sewage can run into any other storm tank. This ensures that the foul first flush cannot pass through to the river if the storm continues for a considerable time. The contents of this tank are returned for full treatment as soon as possible.

In these days when efforts are being made to improve the condition of the rivers, suggestions have been made that storm tanks should have a retention period greater than 2 h at maximum flow; in addition to giving the storm sewage a longer period of sedimentation, this would reduce the incidence of discharges to the river from the storm tanks.

It will be seen that when sewage works are being designed it must always be borne in mind that the detritus tanks and screens must be capable of dealing with the maximum flow which will pass the storm water overflows on the sewers, and the rest of the plant with the maximum flow passing the separation weir on the works.

The Technical Committee on Storm Overflows and the Disposal of Storm Sewage, which was appointed by the Minister of Housing and Local Government in 1955 has recently issued its Final Report[2]. The Committee considered that the custom of expressing the setting of a storm overflow as a multiple of the D.W.F. was basically unsatisfactory and recommended the following formula:

Setting (Q) = D.W.F. + 300 P + 2E (gal/day)

where D.W.F. = dry weather flow of the 'combined' and/or 'partially separate' areas draining to point to overflow (gal/day)
P = population of these areas
E = volume of industrial effluent discharged to the sewers from these areas in 24 hours (gal)

(To obtain Q in m^3, then D.W.F. is expressed in m^3/day and E in m^3)

Where areas drained on the separate system discharge to the combined or partially separate system upstream from the area, Q should be increased by an amount equal to 3 × D.W.F. from the area drained on the separate system.

The Committee considered that low side-weir overflows are inefficient. They suggested that, wherever practical, some form of hydraulic control, such as an orifice or a throttle pipe, should be incorporated as part of all storm-overflow installations. They further suggested that the provision of storage capacity downstream from the overflow could be advantageous as it would delay the discharge to the stream and retain much of the strong first flush.

Scum-boards on storm overflows were considered to be only partially successful in retaining gross solids, and where amenity considerations warrant it, the Committee thought that mechanically raked screens were the only answer to the problem.

Regarding the amount of sewage which should be given *full* treatment at the sewage works it was recommended that this should be arrived at from the following formula:

$3\,PG + I + 3E$ gallons per day

where P = population of whole area draining to works
G = average domestic water consumption (gal/day)
I = infiltration water (gal/head/day)
and E = volume of industrial effluent discharged in 24 hours (gal)

(To obtain the volume to be given full treatment in m^3, then G is expressed in m^3/day, I in m^3/day, and E in m^3).

With regard to storm tanks, the Committee recommended that these should be designed to have a total capacity equal to 15 gal (0·68 m^3) per head of population in the 'combined' or 'partially separate' areas draining to the sewage works.

REFERENCES

1. *Royal Commission on Sewage Disposal*, 1901–15, 9 *Reports* with Appendices; also a *Final Report*, 1915. London; H.M.S.O.
2. *Ministry of Housing and Local Government: Technical Committee on Storm Overflows and the Disposal of Storm Sewage*, Final Report, 1970. London; H.M.S.O.

Chapter 4

PRELIMINARY TREATMENT

INTRODUCTION

When sewage has been delivered, via the sewers, to the treatment works it is given preliminary treatment to remove the larger floating and suspended solid matter, grit and also much of the oil and grease content if present in an appreciable amount. Grease removal is usually carried out after grit and gross solids have been eliminated, but the sequence of grit and gross solids removal tends to vary on different works.

REMOVAL OF GROSS SOLIDS

It is necessary to remove the larger floating and suspended matter, which mainly consists of pieces of wood, rags, paper and faecal matter, in the early stages of sewage treatment, otherwise these would tend to cause damage to pumps and other mechanical equipment and might block pipes, valves and channels. The most common method of removing gross solids is by passing the sewage through screens consisting of metal bars with spaces between, where the solids are retained. There are many types and designs of screens. An alternative method is to pass the sewage through a disintegrator or comminutor in which the solids are cut up into small pieces and then settled out in the primary sedimentation tanks.

SCREENING

The amount of gross solid matter which will be removed from sewage by screens depends on the following factors:

    (a) Nature of the sewage

Preliminary Treatment

(b) Size of the spaces in the screens
(c) Degree to which the solid matter is broken up in the sewerage system.

Screens can be divided into two main categories, namely fixed and moving screens. Fine screens have 13–25 mm ($\frac{1}{2}$–1 in) spaces between the bars, coarse screens usually 51–64 mm (2–2$\frac{1}{2}$ in) but there are some with spaces as large as 150 mm (6 in) The latter type is usually installed as a protection for fine screens against possible damage by large pieces of wood etc. coming down the sewers during periods of intense storms. Screening is sometimes carried out in two stages, first by rough, then by fine screens. There are varying opinions regarding the size of screens. Escritt and Rich[1] suggest that at least 0·09 m^2 (1 ft^2) of submerged screen area should be provided for every 90 000 l (20 000 gal) of D.W.F. The optimum size of screen depends upon conditions such as whether the screens are manually or mechanically raked and how frequently; obviously larger screens will have to be provided when the sewerage system is combined than would be the case where a separate system is used.

The screening chamber should be so designed that the velocity of the flow through it will never be less than 0·5 m/s (1$\frac{1}{2}$ ft/s), to prevent sedimentation. To achieve this, an extra screen is sometimes provided on the larger works, which is only brought into operation in times of heavy rain, by means of automatically controlled motorised penstocks at the inlet and outlet of the screening chamber.

On all but very small works at least two sets of screens in parallel should be provided to allow for maintenance. It is, however, a wise precaution to construct a by-pass on all screening installations, which can be brought into operation in cases of serious blockages of the screens, a major mechanical breakdown or a power failure. At small sewage works, the velocity quoted may be difficult to attain; some unwanted matter may then deposit and provision must be made to remove it.

FIXED SCREENS

Fixed screens usually consist of fixed metal bars with spaces between them through which the sewage passes; the bars often have a tapered cross section with the thickest side facing the flow of sewage to prevent solids from being wedged in the spaces, especially when the screens are being raked.

The simplest type is the hand-raked screen in which the bars are usually set at an angle of 60° to the horizontal, which facilitates raking and presents a greater submerged area to the sewage. Coarse screens are usually of this type, as are some fine screens on the smaller works.

Mechanically raked screens are much to be preferred, as they eliminate the rather unpleasant task of manual raking. They are often set at 90° to the horizontal, although some are set at 60° and others present a concave face to the flow of sewage.

There are various types of raking mechanism. In the older types the rakes were mounted on endless chains which passed over sprocket wheels, one being a driving wheel. The bottom wheels were below water level and exposed to considerable wear. The more modern types have the rakes mounted on arms or suspended chains, thus obviating the necessity for any moving parts of the mechanism to be under water (see *Figure 5*). Hydraulically operated raking mechanisms have also been introduced into the field in recent years.

The raking mechanisms are usually powered by electric motors which can be controlled either manually or automatically. The latter is effected by means of a time switch, or by floats, electrodes or air-bells, which start the motor when the level of sewage, upstream from the screens, rises, due to there being a considerable amount of solid matter on the screens or to a considerable increase in the flow of sewage entering the works.

Some method has to be provided for cleaning the rake tines after they have removed the solid matter from the screens. This is usually done by revolving brushes or star wheels, or by scrapers mounted on spring-loaded arms.

MOVING SCREENS: (A) WIRE ROPE SCREENS

These consist of endless wire ropes moving over rollers and passing through the flow of sewage in the screening chamber. The sewage passes through the spaces between the ropes, the gross solids which are arrested are carried from the screening chamber on the ropes and cleaned off by revolving brushes before the ropes pass back into the screening chamber. This type of screen has the advantage that it dispenses with raking and prevents solids from being wedged in the spaces in the screen.

MOVING SCREENS: (B) DRUM SCREENS

These consist of a revolving drum whose sides are covered by a wire mesh which is partially submerged in the screening chamber.

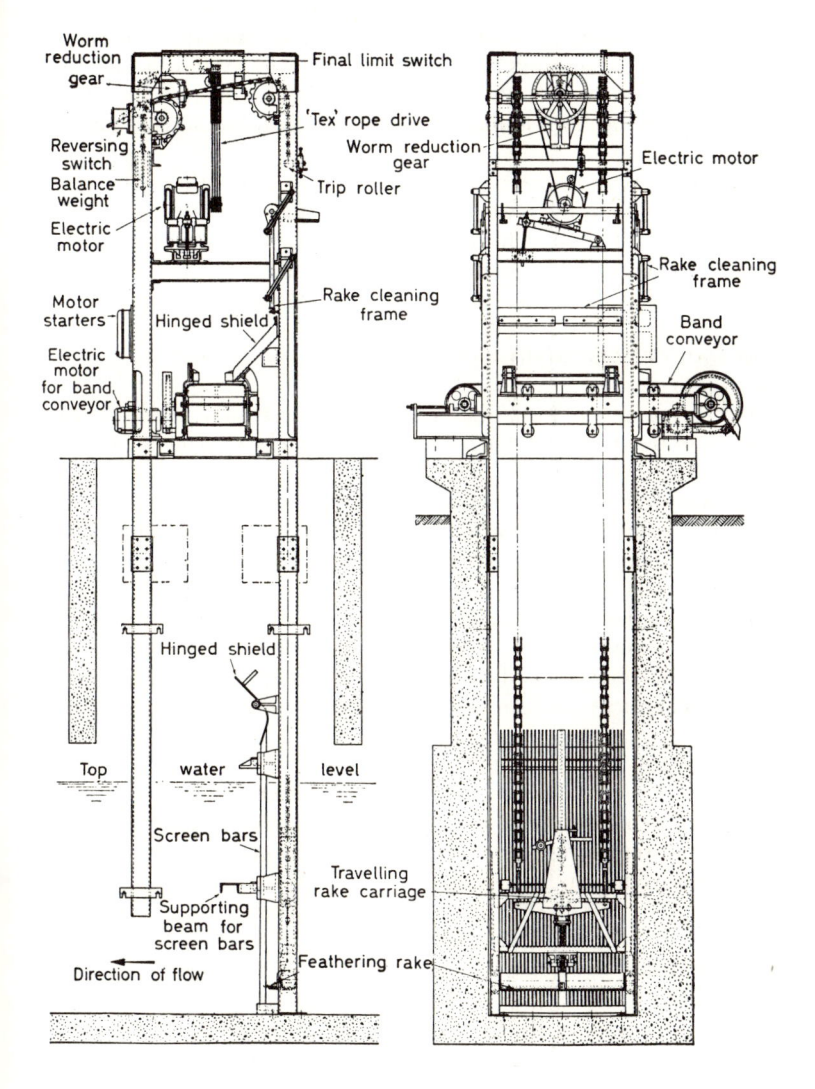

Figure 5. Simplex grab type screen. (By courtesy of Messrs Ames Crosta Mills & Co., Ltd., Heywood)

The sewage enters axially through one end of the drum, the other end being closed, and passes out through the wire mesh. As the drum revolves the solid matter arrested on the inside of the mesh is raised above water level and washed into receiving trays by jets of water from sparge pipes. The wire meshes used are fairly fine, and this type of screen is rarely used in Great Britain for screening crude sewage, as it has been found that the use of such fine screens has few advantages.

A similar type of screen consists of a metal ring with a wire mesh, fixed vertically in the sewage channel in a metal groove and rotated from a central shaft. The solids are washed from the mesh into a trough by jets of water above the sewage level.

DISPOSAL OF SCREENINGS

After being taken from the screens the screenings are as a rule removed by means of barrows, hopper trucks or conveyor belts. The old method of disposal was to bury them in the ground, and this is still favoured at some works today, especially the smaller ones. Burning in an incinerator is a much better method, but nowadays there is a tendency to favour the practice of chopping up the screenings into small pieces in a macerator and then returning them to the sewage channel downstream from the screens; the screenings are usually conveyed from screens to macerator by a water-carriage system. On some works the macerated screenings are returned to the sewage channel upstream from the screens, this is to ensure that the screenings are properly macerated and no large pieces of solid matter pass forward.

Screenings are sometimes washed with water before disposal, the washings being returned to the sewage channel. This reduces the amount of screenings to be disposed of and makes them somewhat less offensive to handle.

COMMINUTORS

Another method of dealing with the gross solid matter, adopted at a number of modern sewage works, consists of cutting the solids into small pieces, without first removing them from the sewage, by a mechanism called a comminutor or macerator. The comminutor made by Messrs. Jones and Attwood Ltd. (see *Figure 6*) consists of a metal drum installed in a chamber within the sewage channel. There are slots in the drum allowing passage to small pieces of solid matter.

Preliminary Treatment 53

The larger pieces are arrested and held by the flow of sewage against the outside of the drum which, driven by an electric motor, rotates. Cutting teeth fixed to the drum pass through fixed combs. The solid matter is thus cut into small pieces. It passes through the slots into

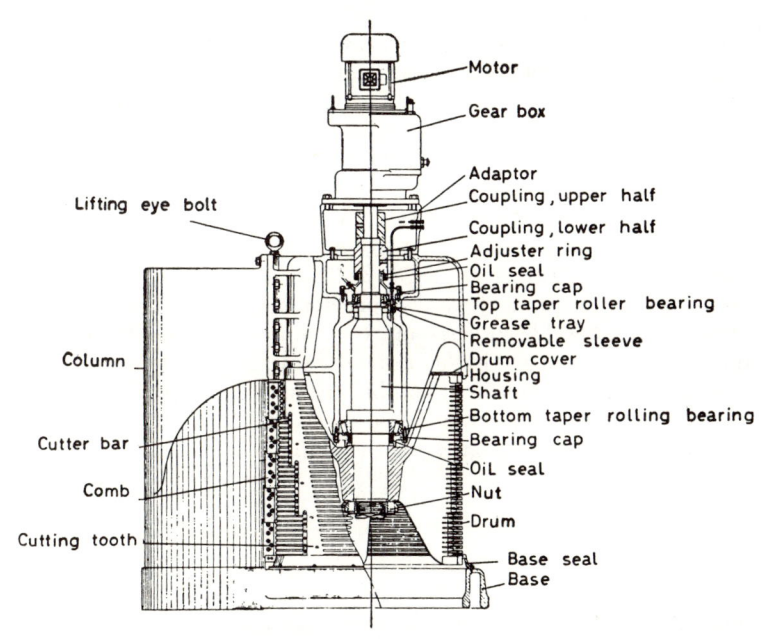

Figure 6. Section of 36T comminutor. (By courtesy of Messrs Jones & Attwood Ltd., Stourbridge)

the drum and then through the bottom of the drum out into the downstream channel via an inverted siphon. It may be seen that this type of equipment involves a loss of head, a point which has to be given careful consideration in cases where the available fall through the works is limited.

As in the case of maceration after screening, this method of dealing with the gross solids places an extra load on the primary sedimentation tanks and increases the sludge disposal problem. Both methods, however, have the advantage that they completely eliminate the handling of offensive screenings.

GRIT REMOVAL

Inorganic gritty matter, or detritus as it is often called, must be removed from sewage in the early stages of treatment, as its abrasive action would otherwise cause damage to pumps and other mechanical equipment. It would also settle in primary sedimentation tanks, making sludge treatment more difficult, and would cause silting in channels and pipe bends. It is removed by sedimentation in grit or detritus tanks. Preferably only grit should settle out in these tanks, and it should be contaminated with as little organic matter as possible—less than about 15 per cent—otherwise it will be offensive and its disposal rendered more difficult. Grit being heavier than the organic solids, careful regulation of the velocity of flow through the grit tanks allows the prevention of most of the organic solids from settling out with it; an alternative method is to wash the grit free of organic matter after it has been removed from the sewage. Opinions as to the amount of inorganic gritty matter present in sewage have varied from 0·019–0·19 $m^3/10^6$ l (3–30 $ft^3/10^6$ gal) of sewage. The fact is that the quantity varies considerably in different sewages. Hence it is advisable to carry out laboratory experiments to determine the amount of grit in the particular sewage to be treated. There is more grit in sewages from combined than from separate systems, and the nature and condition of the streets and roads in the drainage area will have a great bearing on the amount.

GRIT TANKS

The old Ministry of Health recommendation was that there should be at least two grit tanks of a capacity equal to 1 per cent each of the D.W.F. In many of the old works grit tanks often had the form of simple rectangular tanks or channels or even pits, through which the velocity of the flow fluctuated considerably, and in many cases they were too large. The result was that the grit which settled out in these tanks was often contaminated with large amounts of organic matter and was consequently offensive; it was usually removed either manually or by grab-buckets.

To ensure satisfactory settlement of the grit, free from appreciable quantities of organic solids, it is necessary to maintain the velocity of flow through the grit tank at 0·3 m/s (1 ft/s), but at inlet and outlet the velocity should be rather higher 0·45–0·6 m/s ($1\frac{1}{2}$–2 ft/s).

In designing grit tanks for the modern sewage works the accent is placed on maintaining the velocity of flow through the tank at the optimum level, rather than on the capacity of the tank.

There are various types of constant-velocity grit tanks. Many take the form of channels with a cross section to ensure constant velocity. One of the first to be developed was at the Mogden Works[2],

Preliminary Treatment

which is now controlled by the Greater London Council; it has a parabolic curved invert and is controlled by a standing wave flume at the outlet end. Other designs have included modified parabolic cross-sectional and V-shaped channels (*Figure 7*).

There is a variation in nature and size of the particles of inorganic gritty matter present in sewage, but it has been found that the

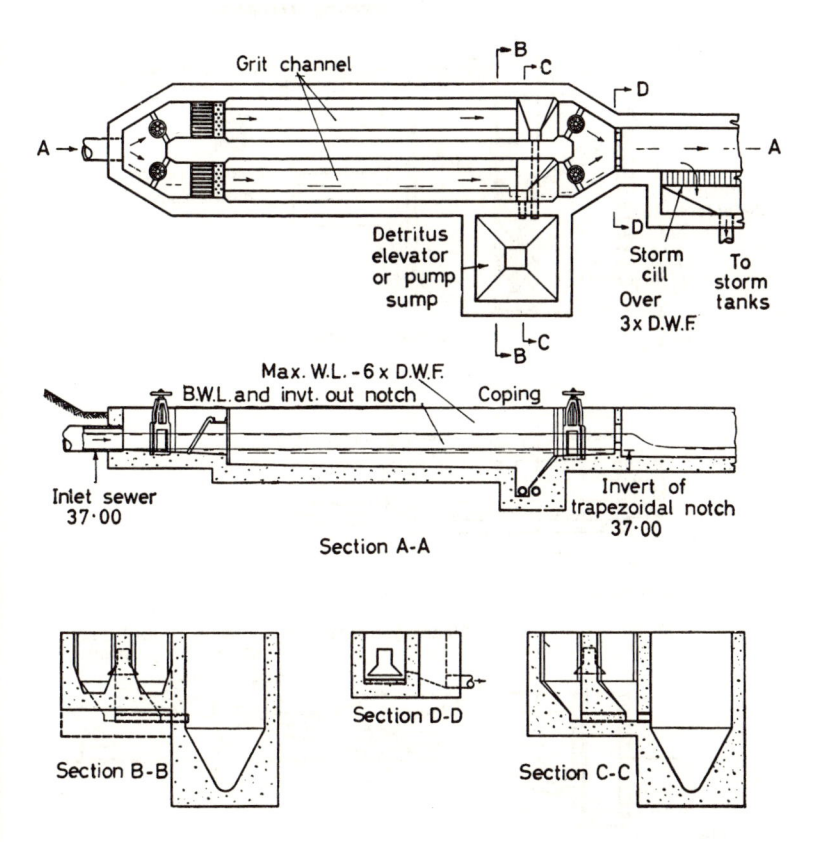

Figure 7. Constant-velocity grit channels. (By courtesy of Adams Hydraulics, York)

majority of the grit will fall 0·3 m (1 ft) in 10 s. Theoretically, if the velocity is 0·3 m/s (1 ft/s), a channel 10 times longer than the maximum depth of sewage should thus settle out most of the grit in the sewage. In practice it is, however, necessary to design the length of channels in the region of 20 times the maximum depth to compensate for turbulence and to allow some of the finer particles to settle.

Grit is generally removed from constant-velocity channels by suction pumps or bucket dredgers mounted on travelling bridges.

Rectangular tanks can be used for efficient grit removal if adapted to maintain a constant flow velocity. One such example is the Dorr Detritor (*Figure 8*), a rectangular tank with an inlet along the whole

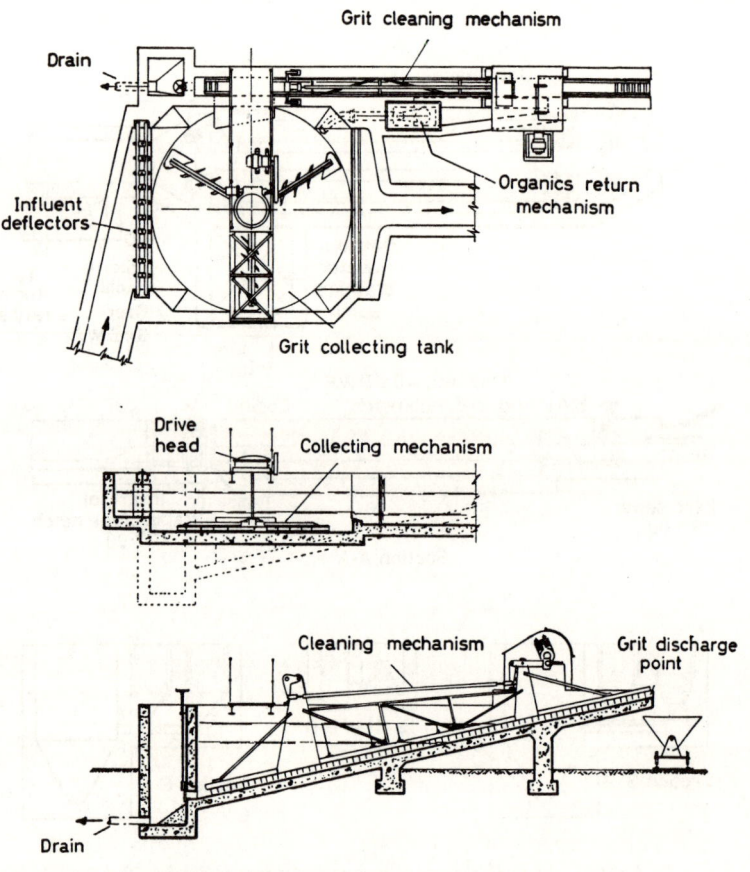

Figure 8. Dorr Detritor. (By courtesy of The Dorr-Oliver Co., Ltd., Croydon)

of one side. A number of gates are fitted along the inlet which can be individually adjusted to maintain constant flow velocity. The sewage flows across the tank and passes over a weir on the opposite side. The grit which settles on the flat bottom of the tank is pushed outwards to the sides of the tank by a rake driven from a central vertical shaft and finally deposited in a pocket. It is then drawn out

Preliminary Treatment 57

of the pocket up an inclined plane by means of a reciprocating rake, and there is a washing action on the grit which has the effect of removing organic matter. Alternatively the grit can be pumped from the pocket up to a classifier in which a vortex effect is set up. The grit with a minimum of water, is discharged into a receiving bay or trailer and the majority of the water, together with the organic matter, is discharged into the channel leading to the primary sedimentation tanks.

Aerated grit chambers are also used successfully to remove grit from sewage. The organic matter is separated from the grit by controlled aeration for a few minutes and fairly large bubbles are used. A very clean grit is obtained and it has been found that the velocity of flow of the sewage in these plants is not critical.

On all but the smaller works it is desirable to have at least two grit removal units, to allow for maintenance to be carried out. Where only one unit is installed it is desirable that a by-pass channel round the unit be provided.

DISPOSAL OF GRIT

On some works it is found desirable to wash the grit to free it from organic matter after it has been removed from the grit tanks; a washer on the lines of the reciprocating rake on the Dorr Detritor, or the classifier described above, can be used. Another grit washer, designed on different lines, is Messrs Ames Crosta Mill's Vortex Grit Separator. The liquor containing grit and some organic matter taken from the grit tanks is fed tangentially into a circular tank with a cone-shaped bottom. A vortex motion is set up in the liquor which causes the heavy grit to collect in the centre of the bottom of the chamber and to fall through a hole into a sump from which it is removed by a bucket elevator. The water containing the organic solids is siphoned off from the centre of the chamber and returned to the sewage flow.

Grit can be disposed of on a tip or can be pumped with water into a lagoon if suitable land is available. If clean, grit can be used for filling in holes, for road making and for gritting roads in frosty weather, and it can also be utilised on sludge-drying beds.

When grit has to be pumped it is desirable that the concentration should be below 5 per cent to prevent excessive wear on the pumps.

REMOVAL OF OIL AND GREASE

Some sewages contain appreciable amounts of oil and grease, especially in drainage areas where there is a high concentration of engineering works, and it is advisable to remove as much of this as possible in the early stages of treatment, in order to prevent any adverse effects on the rest of the plant. The sewage is usually passed through small skimming tanks, where the oil and grease is skimmed off, and this process can be aided by aeration, chlorination or vacuum flotation. Oil and grease can also be removed from sewage by causing them to settle out in the primary sedimentation tanks by the use of chemical coagulants. This method is used to remove grease from sewages containing wool-scouring waste liquors, in which the grease is in an emulsified state and so cannot be removed by ordinary skimming methods.

CHLORINATION

The sewage entering a sewage works is sometimes stale or septic, owing to long flat sewers, extremely warm dry weather, the presence of slime growths or high sulphate content. The septic condition of incoming sewage can give rise to aerial nuisance and may also have an adverse effect on the treatment processes. To reduce smells and to terminate the septic activity, such sewages are sometimes chlorinated as they enter the sewage works. Chlorination of the sewage before primary sedimentation also helps to prevent the settled sludge from becoming septic. This is advantageous, as septic sludge tends to rise, being buoyed up by the gases formed, which has a bad effect on the sedimentation tank effluent. It is often preferable to prevent septicity taking place, and in those areas where the sewers have very little fall or where there are long detention periods in pump wells, chlorine is often introduced at selected points on the sewerage system. This also helps to prevent corrosion of the sewer fabric by gases which are given off from septic sewage. Chlorination also causes reduction in the B.O.D. of sewage, sometimes as much as 15–30 per cent.

The chlorination of sewage poses different problems than does the chlorination of clean water. Wilson[3] pointed out that a small dose of chlorine applied whilst the sewage is in a fresh state will be as effective as at least 10 times the amount applied to sewage which has turned septic. Hypochlorite has been found to be 2–3 times as effective as chlorine. Some workers have found that agitation of the sewage at the time of introducing the chlorine considerably increases its effectiveness.

In the field of sewage treatment it has usually been found that the most convenient way of introducing chlorine is to dissolve the gas in water by using a chlorinator and to dose the chlorine solution into the sewage.

The required chlorine dosage depends upon the nature, condition and strength of the sewage, but dosages between 10 and 25 mg/l have been used with good effects. According to Russell, Munro and Peacock[4], the chlorine demand of sewage in Britain may vary from 5 mg/l (in times of heavy rain) to as much as 30–35 mg/l (in very dry weather). Chlorination should, however, only be resorted to when conditions make it necessary, otherwise desirable biological changes may be suppressed. Chlorine is also used to disinfect sewage effluents (see Chapter 12).

REFERENCES

1. ESCRITT, L. B. and RICH, S. F., *The work of the Sanitary Engineer*, 1949. London; Macdonald & Evans
2. WATSON, D. M., West Middlesex Main Drainage, *J. Instn. Civ. Engrs.* 5 (1936) 463
3. WILSON, H., Some problems in the chlorination of sewage, *J. Proc. Inst. Sew. Purif.* 4 (1950) 433
4. RUSSELL, E. M., MUNRO, E. P. and PEACOCK, T. C., The removal of pathogens during sewage treatment, *Wat. Waste Treat. J.* 8 (1962) 575

Chapter 5

PRIMARY SEDIMENTATION

INTRODUCTION

After gross solid matter, grit and excessive amounts of oil and grease have been removed from the sewage, the next step is to eliminate as much as possible of the remaining suspended solid matter, so as to reduce the strength of the sewage and to make it more amenable to biological oxidation. This is particularly important where the biological purification takes place on percolating filters, because if the liquor dosed onto such filters contains appreciable amounts of suspended solid matter they would soon become clogged. There is a school of thought which maintains that only part, or even none, of the suspended solid matter needs to be removed from the sewage if it is to be treated in an activated sludge plant. In the United States there are activated sludge plants treating sewage from which the suspended solid matter has not been removed. This mode of operation has the disadvantage that it places an extra load on the activated sludge plant and also increases the amount of surplus activated sludge to be disposed of—primary crude sludge is usually considered to be easier to deal with than surplus activated sludge.

Sedimentation is the most practical and economical method of removing suspended solid matter from sewage. This is carried out in primary sedimentation tanks; the solids settle out as a sludge at the bottom of the tank from where it has to be removed; continuous flow tanks are used today. The late Mr. T. Barlow, of Denton, used to say: 'Look after your sedimentation tanks, and the rest of the sewage works will look after itself', and while this may be a sweeping statement, it is a fact that efficient operation and maintenance of the sedimentation tanks pays great dividends in the efficiency of the sewage works as a whole.

Efficient removal of the suspended solid matter by sedimentation

Primary Sedimentation

is one of the cheapest means of reducing the strength of the sewage.

The theoretical basis of sedimentation is illustrated by *Stokes' Law* which shows the factors influencing the falling velocity of small solid particles in a liquid in quiescent state:

$$V = \frac{2g}{9} \cdot \frac{r^2}{\eta} \cdot (d_2 - d_1)$$

where V = rate of fall
r = radius of particle
η = viscosity coefficient of the liquid
d_2 = density of particle
d_1 = density of liquid
g = acceleration due to gravity.

Stokes' Law is truly applicable only to very small particles of solid matter (0·13 mm diam. or less), because for larger particles the effect of the viscosity of the liquid is much reduced.

In a continuous-flow sedimentation tank the velocity of flow of the sewage through the tank has a great effect on the settlement of the solid matter in suspension. The main factors affecting sedimentation in this type of tank are:

(a) Density of particle
(b) Density of liquid
(c) Velocity of flow through the tank
(d) Temperature of liquid—causing short circuiting
(e) Size of particle
(f) Gravitational force.

A great deal of work on the settlement of fairly large gritty particles was carried out by Hazen. He showed that for efficient settlement in a continuous-flow sedimentation tank the velocity of sedimentation of a particle must be such that the particle is able to settle to the bottom of the tank within the distance between the tank's inlet and outlet. Thus a shallow tank is more efficient than a deep tank, provided it is not so shallow as to cause scouring and disturbance of the settled solids at the bottom of the tank. The amount of surface area of liquor in a continuous-flow sedimentation tank is therefore an important factor in its design.

Putting the matter into the simplest terms there are two main forces acting on a particle of suspended solid matter in a sedimentation tank, namely, the velocity of flow through the tank and the gravitational force on the particle. The path of the particle during settlement is roughly their resultant, although these forces themselves are affected by the above-mentioned factors.

Larger particles settle more easily than smaller ones, and stronger sewages have greater relative sedimentation efficiency owing to this fact. The various aids to sedimentation which will be described later rely mainly on building-up of particle size.

The settlement of suspended solid matter in sedimentation tanks has often been considered mainly as physical action, but Stones[1], by experiments using 18 h sedimentation time, shows that if the sewage is sterilised there is a reduction in percentage purification, which suggests that some biological flocculation takes place in the tanks. It would appear that too much stress has been laid on the removal of suspended solids as a criterion of sedimentation efficiency. In view of the biological purification which occurs during tank treatment the detention period in the tanks is a factor to be considered in the design of sedimentation tanks.

SEDIMENTATION TANKS

The Royal Commission on Sewage Disposal, in its 5th Report, made the following recommendations regarding the size of sedimentation tanks for the treatment of sewage:

	Detention
Continuous flow sedimentation tanks	15 h
with chemical coagulation	8 h
Quiescent sedimentation tanks	2 h
with chemical coagulation	2 h
Septic tanks	24 h

The detention period, based on the daily D.W.F. of the sewage, is the theoretical time the sewage would remain in a continuous-flow tank if entering the tank at a uniform rate of flow.

$$\text{Detention time (h)} = \frac{\text{total capacity of tanks (m}^3 \text{ or gal)}}{\text{daily D.W.F. (m}^3 \text{ or gal)}} \times 24$$

For example, for a set of tanks to have a detention time of 8 h their total capacity must be $8/24 \times$ D.W.F. (m^3 or gal).

It has been found that in a well designed continuous-flow sedimentation tank approximately half the suspended solid matter settles out in the first two hours of sedimentation, and there is a tendency these days to consider that the detention periods in the Royal Commission's recommendations are somewhat excessive. Long detention periods can best be justified by the fact that sewages

Primary Sedimentation

in England tend to be colloidal and therefore more difficult to settle. Moreover, in sewages containing a large proportion of industrial waste liquors, a long detention period does assist in the mixing and balancing of the various wastes and helps in preventing shock loads from being passed on to the biological purification plant. Up to 90 per cent of the suspended solids and about 40 per cent of the organic matter should be removed from sewage in an efficient sedimentation tank.

There should be at least two primary sedimentation tanks on a sewage works, the number depending on the amount of sewage to be treated and the size of tank found to be most convenient. The number of tanks should be such that when one tank has to be taken out of action for maintenance or repairs, the reduction in settling capacity does not impair the efficiency of the works. This is specially important where tanks are manually desludged and have to be taken out of action for this purpose. Additional tank capacity is often provided for this reason.

Sedimentation tanks are usually operated in parallel, and it is important to design the feed channels so that the flow of sewage is equally distributed between the tanks. Some of the earlier continuous-flow tanks were operated in series, but this is unsatisfactory, as the velocity of flow through the tanks is greater than it would be through tanks of equal size operated in parallel. In tanks operated in series there are considerable variations in the nature of the sludge which settles out in the individual tanks. Most of the sludge settles out in the first tank, and this has usually to be desludged more frequently than the subsequent tanks.

At a few sewage plants (e.g. Mogden, Coleshill and Luton) sedimentation of sewage is carried out in two stages, using 'primary' and 'secondary'* tanks in series, the former having a much shorter detention time. It is open to question whether this gives better results than the normal procedure, but it could be argued that the use of two-stage sedimentation might be an advantage in helping to balance an industrial sewage.

The design of the inlets and outlets of sedimentation tanks is of prime importance. If badly designed, these can cause excessive flow velocities in sections of the tanks, and also short-circuiting and channelling, with consequent reduction of settling efficiency. In designing a tank inlet the object is to dissipate the flow velocity built up in the feed channel and to distribute the flow evenly throughout

---

* These are not to be confused with the final sedimentation tanks following filters or activated sludge plants which are often referred to as 'secondary sedimentation tanks'.

the tank. To achieve this, baffles are often fitted to individual inlets, or a single baffle board in the tank itself, and often the inlets are situated below the top water level in the tank. These measures also prevent the warmer incoming liquor from passing over the surface of the colder liquor in the tanks. Outlets usually take the form of a long weir, thus allowing the settled sewage to be drawn off as a thin film. The use of notched outlet weirs has become popular in recent years.

During sedimentation in tanks a scum usually forms on the surface of the sewage, consisting of floating matter and, in some cases, oil and grease. To prevent this passing through the outlet a scum board is placed in front of the outlet. The top of the board is usually 75–150 mm (3–6 in) above, the bottom 300–375 mm (12–15 in) below the surface of the sewage. The scum board should not be placed too near to the outlet, as this would increase the velocity of flow there. Scum removed is usually mixed with the sludge for disposal.

The settled sludge has to be removed from the sedimentation tanks at frequent intervals, otherwise anaerobic bacterial action would set in, with the result that the sludge would become septic and, buoyed up by the gases formed, would begin to rise to the surface. This would increase the suspended solids and organic content of the tank effluent, thus putting an extra load on the biological purification section of the works. Sludge is removed from sedimentation tanks either by decanting off the supernatant water and then manually pushing the sludge to the sludge outlet by squeegees, or by scraping it into hoppers by mechanical means and then removing it by pumps or by hydrostatic head; exceptions are hopper-bottomed tanks where the sides of the hoppers are built at an angle steep enough to allow the sludge to fall to the bottom from where it can be removed.

At present, manually desludged tanks are very rarely installed except on small works. Although they are cheaper to construct, they have the disadvantages that desluding is unpleasant and that manual labour is expensive and difficult to obtain. In addition a tank has to be taken out of action for desludging with a consequent reduction of sedimentation capacity, and extra tanks have often to be provided for this purpose. On the other hand, hopper-bottomed tanks and those fitted with mechanical scrapers can be desludged while in operation. It is rarely practicable to manually desludge a tank more frequently than once per week, but from the other types of tanks sludge can be drawn at least once a day, so that there is no reduction of sedimentation capacity due to accumulated sludge and less risk of it becoming septic in the tanks. The reduction in detention time of a manually desludged tank due to accumulated sludge must be taken into consideration when designing the tank.

Primary Sedimentation

TYPES OF SEDIMENTATION TANKS

Sedimentation tanks usually fall into one of the following classes.
(a) *Fill-and-draw quiescent sedimentation tanks*—The simplest form of sedimentation tank is the fill-and-draw tank in which sewage can be given quiescent sedimentation. This consists of a simple rectangular tank with inlets, a decanting arm with a scum board fitted round the mouth and a sludge outlet valve. There is usually also an overflow outlet which comes into operation if, owing to lack of attention, the inlets are not closed when the tank is full. The mode of operation is for a tank to be filled with sewage and then allowed to stand for at least 2 h at D.W.F. The settled sewage is drawn off via the decanting arm and the sludge removed from the tank. The tank is then ready for filling up again. It can be seen that at least four tanks are necessary, and for efficient operation it is necessary to have even more to cope with maintenance and increased flows in wet weather. This type of tank has never been widely adopted, as it requires almost constant attention. It is costly in labour and does

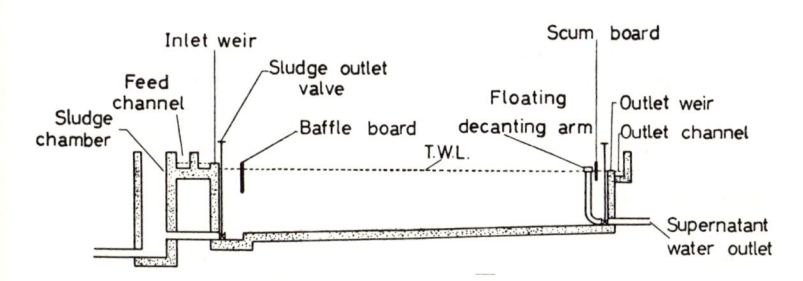

Figure 9. Horizontal flow sedimentation tank

not allow for much flexibility of operation. Moreover, there is a much greater head loss through a tank of this type than through a continuous flow tank. This type of tank would not generally be considered for any new sewage works today and is mainly used for the treatment of trade wastes which vary in character and have to be dealt with in 'batches'. It is usually designed so that if the inlet is not closed when the tank is full, it will operate with reasonable efficiency as a continuous flow tank.

(b) *Horizontal Flow Tanks*—These usually consist of rectangular tanks through which the sewage is continually flowing from an inlet at one end to an outlet at the other (*Figure 9*). The detention period in the tank and velocity of flow through it are such that the bulk of the suspended solid matter settles to the bottom of the tank from where it is subsequently removed. The tanks are usually constructed with a length-to-breadth ratio of 3 or 4 to 1—a long narrow tank would increase the velocity of flow, with a consequent adverse effect on sedimentation efficiency. The average depth is 1·8–3 m (6–10 ft) and the usual detention capacity 10–15 h at D.W.F., although there is a school of thought (*see* p. 62) which considers that lower detention capacities can be used to obtain efficient sedimentation.

Horizontal flow tanks can be sub-divided into 1, manually and 2, mechanically desludged tanks.

1. *Manually desludged tanks*—The floor of this type of tank slopes down to the inlet end where the sludge outlet valve is located, as the bulk of the sludge, consisting of the heavier and larger particles, settles out near the inlet end. The slope of the floor should not be greater than will allow workmen to walk upon it without difficulty; a slope of 1 in 25 is usually suitable. The inlet consists of either a weir across the whole width of the tank or baffled ports. In front of the inlet is often fitted a baffle board across the tank. At the opposite end is the outlet which consists of a weir across the width of the tank, in front of which is fitted a scum board. A decanting valve, consisting of a floating arm on a swivel or a telescopic pipe, is located near the outlet end. To desludge a tank of this type, the inlet is first closed and the sewage in the tank allowed to settle. The supernatant liquor is decanted off and should be pumped back to the inlets of the other tanks. On some works the supernatant liquor is run directly to the biological treatment plant, but this is not a satisfactory practice, especially in the case of percolating filter plants, as they have been known to become clogged owing to this procedure. The sludge is then pushed down to the sludge outlet by men with squeegees; this method is improved at some works by using a small tractor fitted with a scraper board to push the sludge. The tractor enters the tank either by means of a crane or a ramp. This cuts down labour costs and makes the work less unpleasant.

The sludge from this type of tank usually has a water content of 92–94 per cent (i.e. the total dry solids content is 6–8 per cent).

The scum held back by the scum board has to be skimmed off manually.

2. *Mechanically desludged tanks*—These tanks are similar in shape to those which are manually desludged, but the floors are virtually flat with inverted, pyramid-shaped hoppers at the inlet end. The Mieder type sludge scraper mechanism is the one most commonly used in England. It consists of a power-driven bridge which spans the width of the tank and runs on rails along the top of the tank walls. A scraper, suspended from the bridge by arms, is driven along the floor of the tank and thus pushes the sludge into the hoppers

Primary Sedimentation

from which it is removed either by hydrostatic head or by pumping. The scraper can also be used to remove scum from the surface of the sewage in the tank by being raised to the surface on the bridge's return journey; it pushes the scum into a trough from where it runs into the sludge well. Alternatively the bridge can be fitted with a special skimmer for pushing the scum into a trough.

One mechanical scraper of this type can be used for a number of tanks.

There is another type of mechanical scraper, known as a 'link-belt' or 'flight-conveyor' type. This consists of a number of scrapers fixed on two endless chains and mounted on and driven by sprocket wheels. The scrapers move along the floor of the tank pushing the settled sludge into the hoppers. They then rise to the surface and return to the other end of the tank, pushing scum into a collecting trough as they do so. This type of mechanism has been very rarely used in the British Isles: there is a limit to the length of scraper so that in a very wide tank a number of them would have to be used side by side; moreover, such a scraper cannot be moved from one tank to another.

The water content of sludge from a rectangular tank fitted with mechanical scrapers is about 92 per cent. The speed at which the scrapers pass through the tanks must be slow enough to prevent excessive disturbance of the sludge or turbulence in the tanks; Babbitt[2] states the usual speed to be about 5 mm/s (1 ft/min) in the U.S.A.

In another type of rectangular horizontal flow tank, the floor is composed of a number of hoppers into which the sludge settles and from where it is removed by pumps or hydrostatic head. This type of tank is mainly used for secondary sedimentation following percolating filters.

Rectangular horizontal flow tanks have the advantage that they can be built in batteries with common walls, and one Mieder type scraper can be used for a number of tanks. They have, however, usually to be built with a greater detention capacity than upward flow or radial flow tanks.

Upward flow tanks—In this type of tank the sewage enters at a point below top water level but above maximum sludge level. The direction of flow is upwards and the velocity of flow causes the particles of solid matter to rise, but their weight reduces their velocity until they are vitually stationary. A sludge blanket forms across the tank which acts as a filter, and the particles form larger aggregates and sink to the bottom of the tank. Upward flow velocity is a very important factor in the performance of this type of tank; a velocity of 1·2–1·8 m/h (4–6 ft/h) at maximum flow is found to be most suitable for the sedimentation of sewage. The surface area of the tank is the main operative factor in determining the rate of upward flow.

The early types of upward flow tanks consisted of deep circular tanks into which the sewage entered via ports in the walls near the bottom and the settled sewage overflowed into channel weirs across the top. The settled sludge on the flat bottoms of the tanks was pushed into central hoppers by means of scrapers manually rotated by vertical shafts passing through the centre of the tanks.

The modern upward flow tank is either square or circular in plan, with an inverted cone or pyramid-shaped hopper bottom. It has a centre feed, but as it is important to prevent turbulence or channelling, the inlet has to be well baffled and is surrounded by a box or cylinder, the open bottom of which is well below water level (see *Figure 10*). The settled sewage passes over an outlet weir running round the perimeter of the top of the tank. The sludge settles into

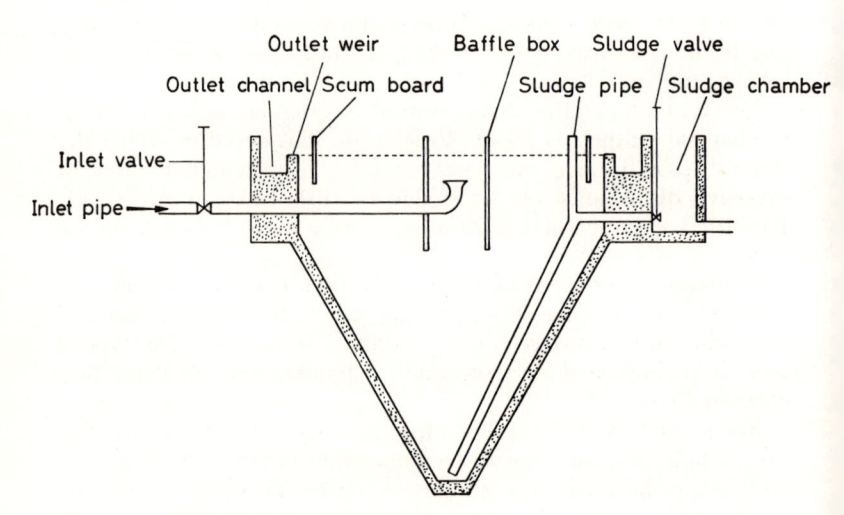

Figure 10. Hopper-bottomed upward flow tank

the hopper, the sides of which should make an angle of not less than 60° with the horizontal to allow the sludge to fall to the bottom of the hopper instead of lodging on the sides. The sludge is usually drawn from the hopper by hydrostatic head, by means of a pipe which passes from the bottom of the tank to an open end above top water level (for cleaning purposes), and with a branch draw-off valve below water level which when opened allows sludge to be drawn from the hopper. The pressure is created by the difference in head between the top water level and the draw-off valve; this

Primary Sedimentation

should not be less than 1·2 m (4 ft) when dealing with sewage sludge.

The sludge removed from this type of upward flow tank usually has a water content of about 96 per cent.

Square upward flow tanks have the advantage that they can be built in batteries with common walls; for these hopper-bottomed tanks, however, large excavations have usually to be made to accommodate the hoppers which have to be made deep enough to obtain a minimum 60° slope on their sides. The greater the width of the tank the deeper it has to be, but in larger tanks this difficulty can be overcome to some extent by constructing four hoppers in the bottom instead of one.

Radial flow tanks—These tanks are circular in plan and the sewage enters at the centre, usually into a vertical cylindrical drum or chamber, from where it passes through the bottom and/or via ports into the body of the tank. It then flows towards the perimeter of the tank, round the whole of which is an outlet weir and channel into which the settled sewage overflows (*Figure 11*). There is often

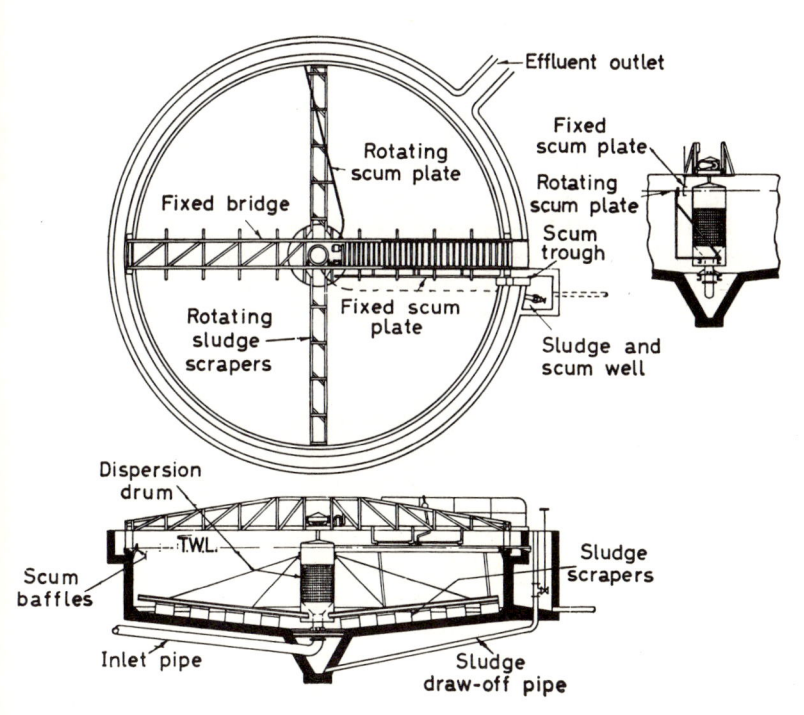

Figure 11. Simplon settling tank. Radial flow. (By courtesy of Messrs Ames Crosta Mills & Co., Ltd., Heywood)

a scum board fitted in front of the overflow weir. In this type of tank the velocity of flow is greatest near the centre where the larger and heavier particles settle out, and much less near the perimeter, thus allowing the smaller and lighter particles to settle. Although called a radial flow tank there is also an upward-flow effect. Primary sedimentation tanks of this type are usually designed for 6–8 h detention time at D.W.F.

There are a number of patented designs for this type of tank, all of which, having floors which slope down to the centre of the tank in varying degrees, are cleaned by mechanical sludge scrapers which sweep the sludge into hoppers, usually situated at the centre of the tank. Some scrapers are driven from the centre shaft, others are fixed to a travelling bridge which is pivoted at the centre of the tank and runs on a rail fixed on top of the tank wall. In addition some have skimmers which, dipping just below the surface of the water, move round the tank and push the scum into the receiving troughs. The hoppers should be large enough to hold at least one day's accumulation of sludge, which is removed by hydrostatic head or pumping. The peripheral speed of scrapers should not exceed 25 mm/s (5 ft/min).

The sludge removed from this type of tank usually has a water content of about 95 per cent.

By reason of their shape these tanks cannot be built in batteries with common walls, but they are much shallower than upward flow tanks with a corresponding saving in excavation costs.

In all continuous flow tanks there is, in varying degrees, some upward-flow effect; for this reason, as well as those previously mentioned, the surface area is a very important factor in their design. Therefore, their operational performance is sometimes expressed in the following terms:

surface loading (m^3/m^2 day of tank surface area)
(Metric) $= \dfrac{\text{24 h amount of sewage } (m^3)}{\text{tank surface area } (m^2)}$

surface loading (gal/ft^2 day of tank surface area)*
(British) $= \dfrac{\text{24 h amount of sewage (gal)}}{\text{tank surface area } (ft^2)}$

* *Average figures for*	l/m^2 day	gal/ft^2 day
Primary tanks	9 800–14 700	200–300
Activated sludge final tanks	39 200–49 000	800–1 000
Humus tanks	14 700–19 600	300–400

Primary Sedimentation

overflow rate (relationship of surface loading to length of outlet weir)
(Metric) $= \dfrac{\text{Surface loading} \times \text{tank surface area (m}^2\text{)}}{\text{length of outlet weir (m)}}$

overflow rate (relationship of surface loading to length of outlet weir)
(British) $= \dfrac{\text{surface loading} \times \text{tank surface area (ft}^2\text{)}}{\text{length of outlet weir (ft)}}$

It should always be remembered that primary sedimentation tanks have usually to deal with flows up to $3 \times$ D.W.F., which must be taken into consideration in their design.

AIDS TO SEDIMENTATION

Sewage contains colloidal matter and very finely divided suspended solids (pseudo-colloidal matter), of which very little is removed by plain sedimentation. To remove some of this during sedimentation, it is necessary to cause the fine suspended matter to combine to form large particles which will settle more easily, or to form insoluble flocculent precipitates which will absorb some of the colloidal and finely divided matter. There are various methods of attaining this, the main types being mechanical flocculation and chemical coagulation.

MECHANICAL FLOCCULATION

If sewage is stirred gently the finely divided matter coalesces into larger particles which settle more easily. To carry this out on a sewage works the sewage is passed through a tank with 20–30 min detention time at D.W.F., in which are slowly rotating paddles of an optimum peripheral speed of about 0·43 m/s (1·4 ft/s); excessive paddle speeds will tend to break up the flocs. The sewage then passes into the sedimentation tanks.

The flocculating chamber is sometimes constructed as part of a sedimentation tank, as in Messrs. Dorr Oliver's Clariflocculator (*Figure 12*). This is a circular radial flow tank with mechanical sludge scrapers, which has a circular chamber in the centre fitted with paddles mounted on a power-driven vertical shaft. The sewage passes through this flocculating chamber before passing into the outer chamber where sedimentation takes place. Hurley and Lester[4] showed that a Clariflocculator can produce an effluent about 20 per

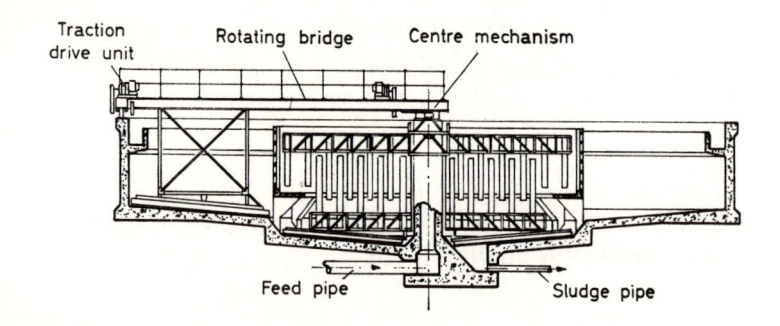

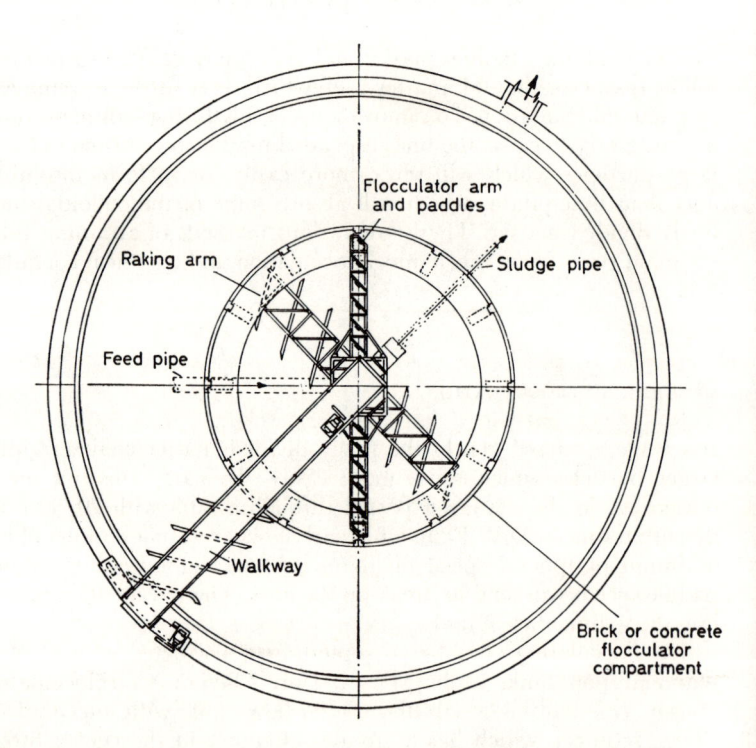

Figure 12. Dorr Clariflocculator. (By courtesy of The Dorr-Oliver Co., Ltd., Croydon)

Primary Sedimentation

cent better than that obtained by plain sedimentation, at very little extra cost.

Improved results have been obtained from flocculation by returning sludge from the sedimentation tanks back to the flocculating tank. Marsden[5] carried out large-scale flocculation experiments at the Bury Corporation's Blackford Bridge Sewage Works. At D.W.F. the sewage had 3 h each detention time in the flocculation and the sedimentation tank. The flocculation mechanism consisted of a large high-intensity Simplex aeration cone revolving at 38 rev/min, and sludge from the sedimentation tank was returned to the flocculation tank. Compared with plain sedimentation the percentage purification, based on the 4 h permanganate value, increased by almost 100 per cent. The effect was interpreted as a form of bio-flocculation.

Some experimental work has been carried out on the use of ultrasonic waves as an aid to sedimentation, but the results are not so far conclusive.

CHEMICAL COAGULATION

The addition of certain chemicals to sewage causes the formation of an insoluble flocculent precipitate which absorbs and carries down some of the suspended and colloidal matter. With many of these chemicals the pH value of the sewage is of great importance. Some of them are used in conjunction with one another. It is usually desirable to add chemicals in the form of a solution or suspension in water, which requires mixing and dosing gear. The addition must be well mixed with the sewage; this can be carried out in baffled channels. Moreover, addition of chemicals is often followed by mechanical flocculation before sedimentation, which improves the result.

The following chemicals are used for chemical coagulation.

Hydrated lime $[Ca(OH)_2]$—It is usually maintained that hydrated lime reacts with all the free carbon dioxide and calcium bicarbonate in the sewage to produce an insoluble precipitate of calcium carbonate, which acts as flocculating agent carrying down much of the impurity.

$$Ca(OH)_2 + H_2CO_3 = \underset{\substack{\text{insoluble}\\\text{calcium}\\\text{carbonate}}}{CaCO_3} + 2H_2O$$

$$Ca(OH)_2 + Ca(HCO_3)_2 = 2CaCO_3 + 2H_2O$$

Stones[6], however, carried out experiments using lime as a coagulant and put forward the suggestion that in the precipitation of negatively charged sewage colloids lime behaves as if it were a positively charged calcium hydroxide sol.

A large excess dosage of lime has a deleterious effect, since it dissolves or disperses some of the organic suspended matter and so gives rise to an inferior tank effluent.

Iron and aluminium salts are also used as chemical coagulants. The positively charged metallic ions in these electrolytes cause coagulation of the negatively charged sewage colloids. The coagulating power of electrolytes increases with increase in valency of the positively charged ion, e.g. Fe''' and Al''' are superior to Ca''.

Aluminium sulphate $[Al_2(SO_4)_3]$—In Britain the substance is generally used in commercial form, Alumino-ferric, which contains a small amount of ferric sulphate. Aluminium sulphate reacts with the calcium bicarbonate in sewage to give an insoluble precipitate of aluminium hydroxide as flocculating agent.

$$Al_2(SO_4)_3 + 3Ca(HCO_3)_2 = 3CaSO_4 + \underset{\substack{\text{insoluble} \\ \text{aluminium} \\ \text{hydroxide}}}{2Al(OH)_3} + 6CO_2$$

The optimum pH range for efficient coagulation with aluminium sulphate is in most cases 5·5–7·0; in some sewages whose pH is affected by trade waste discharges, it may be necessary to adjust the pH with lime or sulphuric acid.

The amounts of aluminium sulphate required range from 50 to 150 mg/l, depending on the nature of the sewage to be treated. Excessive dosages of aluminium sulphate will cause dissolution of the aluminium hydroxide.

Iron salts—The iron salts which can be used are ferrous sulphate or copperas ($FeSO_4$), ferric chloride ($FeCl_3$) and chlorinated copperas ($FeClSO_4$). The commonest and cheapest is copperas, generally used in conjunction with lime. Ferrous bicarbonate, formed first, is by the addition of lime converted into ferrous hydroxide, and finally by the oxygen present in the sewage, into insoluble ferric hydroxide which acts as an efficient coagulant. The final stage can be assisted by air blowing.

$$FeSO_4 \cdot 7H_2O + Ca(HCO_3)_2 = \underset{\substack{\text{ferrous} \\ \text{bicarbonate}}}{Fe(HCO_3)_2} + CaSO_4 + 7H_2O$$

$$Fe(HCO_3)_2 + 2Ca(OH)_2 = Fe(OH)_2 + 2CaCO_3 + 2H_2O$$
ferrous hydroxide

$$4Fe(OH)_2 + O_2 + 2H_2O = 4Fe(OH)_3$$
insoluble ferric hydroxide

For efficient coagulation the pH value should be within the range 8·5–9·0.

Ferric chloride containing trivalent iron is a better coagulant than ferrous sulphate which contains divalent iron, but it is expensive, not available in large quantities in Great Britain and therefore very rarely used. Chlorinated copperas, made on the site by treating copperas with chlorine, is used as a substitute and has given good results in many cases.

With iron salts there is no danger of dissolving the ferric hydroxide by excessive use of the chemical, as ferric hydroxide is not normally amphoteric. Iron salts, however, have the disadvantage that they often give rise to the presence of iron in the sedimentation tank effluent, and at times in the final effluent. In addition the pH value of the sedimentation tank effluent has sometimes to be reduced with acid prior to biological treatment.

Acids—Acids are used in special cases, e.g. where trade waste discharges have resulted in a sewage with high pH value or large amounts of grease, especially in areas where a great deal of wool scouring is carried on, as in the West Riding of Yorkshire. There the sewage is treated with sulphuric acid to precipitate the fatty acids, which settle out in the sedimentation tanks together with the suspended solid matter in the sewage. The sedimentation tank effluent has sometimes to be treated with alkali (e.g. an alkaline sewage) to raise its pH value before it receives biological treatment.

Stones[7] has shown that too small a dose of chemical coagulant can actually reduce the amount of purification effected in a sedimentation tank, as it has the effect of sterilising the sewage and thus precludes any biological action in the tank.

Although chemical coagulation plus sedimentation can result in a reduction of up to 60 per cent in the B.O.D. figure of sewage, it is not extensively used in Great Britain at the present time. In the early days of sewage treatment this was in many cases the only treatment given to sewage before discharging it into a watercourse. Chemicals are expensive, however, and owing to the precipitation of large amounts of inorganic solids, coagulation produces much (up to 50 per cent) more sludge than does plain sedimentation, and so increases the sludge disposal problem. Chemical coagulation processes cannot deal with substances present in sewage in true solution, unless there is a chemical reaction between the added chemical and the substance in solution and true chemical precipitation results. Flocculation and chemical coagulation can be used to increase the efficiency of existing sedimentation tanks which are overloaded, or where seasonal variations occur in the strength and volume of the sewage (e.g. at holiday resorts). As in the treatment of sewages containing wool-scouring wastes, chemical coagulation can be successfully used to assist in the treatment of sewages containing large amounts of trade waste.

Polyelectrolytes [7a, 7b, 7c]—The use of polyelectrolytes as a means of assisting flocculation has become more prominent in recent years. These synthetic substances fall into two main categories, namely, those which have a long chain molecular structure and have the property to be adsorbed on two or more particles, thus drawing the particles together, and those which reduce the charge on the particles and consequently reduce the repulsive power of the like charges on the particles.

Polyelectrolytes tend to be selective in their effects and as sewages tend to vary in character care has to be taken in selecting the most appropriate type of polyeletrolyte.

Polyelectrolytes are derived from such substances as acrylic and alginic acids. For example, use has been made of polymers of cyanamide, acrylic and methacrylic acids or their derivatives, and hydrolysed polymers of acrylonitrile or acrylamide with molecular weights in the range 10 000–100 000. They are used in dilute aqueous solution and usually only very small amounts are required (sometimes as little as 0·25 mg/l) but they are expensive compared with ordinary chemical coagulants.

PRE-AERATION PRIOR TO SEDIMENTATION

Pre-areation of sewage for short periods (10 min or more) prior to sedimentation is used as a means of assisting treatment (especially when stale or septic). Aeration not only assists in the removal of entrained gases (H_2S and CO_2) from sludge particles and assists flocculation, but also aids in the separation and removal of oils and grease.

Several plants in the U.S.A. are now employing pre-aeration and in some cases up to 8 per cent improvement is claimed in the reduction of B.O.D. and suspended solids in the primary sedimentation tanks. In this country Morton and Thomas[8] showed that the Davyhulme

Sewage Works of Manchester Corporation pre-aeration of sewage, mixed with surplus activated sludge, resulted in an improvement in the settling characteristics of the sewage and in the performance of the primary sedimentation tanks, but it had no appreciable effect on the oxygen demand. Stones[9] found that pre-aeration of Salford's industrial sewage resulted in increased removal of suspended solids, but that pre-aeration effected very little improvement on purely domestic sewage.

SEPTIC TANKS

Septic tanks are, in effect, horizontal flow sedimentation tanks in which the sludge settling at the bottom is allowed to digest and liquify by anaerobic bacterial action. This conversion of the solids to liquids and gases is, however, only partially accomplished, and the sludge has, therefore, to be removed from time to time, though a small portion is always left in the tank for seeding purposes. A septic tank differs from an ordinary sedimentation tank mainly in the length of detention of the deposited sludge. The original septic tank of Cameron, constructed at Exeter in 1895, was covered, but the Royal Commission on Sewage Disposal decided that this was unnecessary, since a thick scum forms on the surface of the tank which effectively prevents access of oxygen from the air and allows the anaerobic action to proceed. Nevertheless, from a public safety point of view it is desirable that such tanks should be covered, with adequate ventilation for the escape of gases, or securely fenced round. Septic tanks have the advantage of needing little attention other than the occasional removal of digested sludge from the bottom, and are therefore favoured for very small communities such as a few isolated houses.

The effluent produced, which generally has an objectionable smell, should not be discharged to a stream without further treatment either on land or on a percolating filter. At times, especially in warm weather, the effluent from a septic tank may be even worse than the incoming sewage and may contain more suspended matter due to the buoying up of sludge particles by the gases evolved. This disadvantage can be overcome by using an Imhoff tank, which is really a two-compartment tank in which sedimentation of the sewage is carried out in the upper tank, and the sludge falls into the lower tank where digestion takes place. A few tanks of this kind are still in use today in England, Germany and America. For large sewage works, however, septic tanks are no longer favoured. According to Peel[10], septic tanks, even if operating satisfactorily usually yield effluents having a B.O.D. and suspended solids content over the range 200–400 mg/l. The detention period is normally 1–3 days and the frequency of de-sludging is about at 6 monthly intervals.

The size of septic tank recommended by the British Standards Institution for plants serving approximately 60 persons[11] is given by the formula:

$$C = 30N + 400$$

where C = capacity (gal)
and N = number of persons in full-time residence.

Sewage containing synthetic detergents (50 mg/l, expressed as Manoxol O.T.) can be satisfactorily treated in a septic tank[12]. This amount of detergent is much above the average in sewage.

REFERENCES

1. STONES, T., The settlement of sewage, *J. Proc. Inst. Sew. Purif.* 4 (1956) 349
2. BABBITT, H. E., *Sewerage and Sewage Treatment*, 7th Ed., 1953. New York; Wiley, London; Chapman & Hall
3. *Ministry of Housing and Local Government, Report of an informal working party on the Treatment and Disposal of sewage sludge*, 1954. London; H.M.S.O.
4. HURLEY, J. and LESTER, W. F., Mechanical flocculation in sewage purification, *J. Proc. Inst. Sew. Purif.* 2 (1949) 193
5. MARSDEN, G. R. C., Investigation into the primary treatment processes at Bury Sewage Works, *J. Proc. Inst. Sew. Purif.* 1 (1954) 61
6. STONES, T., Studies on lime precipitation at Salford, *J. Proc. Inst. Sew. Purif.* 4 (1954) 395
7. STONES, T., Influence of toxicity on chemical precipitation, *J. Proc. Inst. Sew. Purif.* 4 (1952) 417
7a. ANON., A new flocculating agent [Separan 2610*]. *Wat. Sanit. Engr.* 6 (1956) 161; also Dow Chemical Co., Midland, Michigan, 1961
7b. *Cyanamide flocculants*. Cyanamide of Gt. Britain, Aldwych, London, 1959
7c. ANON., New coagulant aids remove suspended particles. *Rohm and Haas Reporter, Pa.*, 21 (1963) No. 3, 3. Available from Lennig Chemicals Ltd., Bedford Row, London W.C.1
8. MORTON, A. Y. and THOMAS, A., Pre-aeration channel performance at Manchester, *J. Proc. Inst. Sew. Purif.* 6 (1962) 565
9. STONES, T., Pre-aeration as an adjunct to short period settlement of sewage, *J. Proc. Inst. Sew. Purif.* 1 (1958) 79
10. PEEL, C., Providing lay-by lavatories on sites without main drainage, *Munic. Engng., Lond.* 143 (1966) 1219, 1221–2
11. British Standards Institution, Code of Practice CP 302, 100 (1956) London
12. TRUESDALE, G. A., and MANN, H. T., Synthetic detergents and septic tanks, *Surveyor*, 9th March 1968

---

* Now called Separan NP10.

Chapter 6

AEROBIC BIOLOGICAL TREATMENT

INTRODUCTION

When the famous 5th Report of the Royal Commission on Sewage Disposal[1] appeared in 1908, discussing in detail methods then in use for the treatment and disposal of sewage, the normal method of dealing with sewage was land treatment. This involved either irrigation over a large area of land ('broad irrigation') or filtration through a porous loamy soil ('intermittent downward filtration') which resulted in aerobic oxidation of the sewage to harmless end products by soil micro-organisms.

Septic tanks in those days also played a not unimportant part in sewage treatment, especially in the smaller sewage works, but for the larger plants, contact beds and percolating filters were coming into use.

Contact beds are large tanks containing graded filtering media used for treating sewage by the fill-and-draw method. Settled sewage is allowed to remain in the bed for about 2 h, after which the bed is emptied and rested whilst the effluent is allowed to pass to a second bed if necessary. Purification is similar to that occurring during land treatment. Nitrates formed in the interstices of the media whilst the bed is standing empty assist in the oxidation of the sewage when the bed is filled. The contact bed is virtually obsolete to-day, since percolating filters or the activated sludge process can deal with larger volumes of sewage and yield much better effluents.

Except for use as a final polishing device for sewage effluents (see Chapter 7), land treatment is also now virtually obsolete because of the large areas of land required, the production of unpleasant smells and the tendency of the land to become septic or 'sewage sick'.

The recommendations of the Royal Commission on the design and operation of percolating filters led in the years following the publication of the 5th Report to the construction of many new percolating filter installations in the British Isles and abroad. With the discovery by Arden and Lockett[2] in 1913 of the activated sludge process, a new and outstanding aerobic biological process rivalling the percolating filter was made available for the purification of sewage, and these now constitute the most important modern sewage treatment processes. During the past 30 years, both these processes have been further developed and improved in performance and efficiency, so that by using these methods either alone or in combination, or in conjunction with certain polishing devices (see Chapter 7), it is technically possible to obtain a final sewage effluent of almost any desired degree of purity.

In the biological purification of sewage and other organic waste waters in the presence of sufficient air, bacteria and other microorganisms bring about a number of changes in the following order:

1. *Coagulation stage,* i.e. coagulation and flocculation of colloids and pseudo-colloids
2. *Oxidation stage,* i.e. oxidation of carbonaceous matter to carbon dioxide (p. 13)
3. *Nitrification stage,* i.e. oxidation of ammonia, derived from the breakdown of nitrogenous organic matter, to nitrite and eventually nitrate (p. 14)

The third or nitrification stage takes place fairly completely in well operated low-rate percolating filters. In activated-sludge plants it has, in the past, generally been considered uneconomic to nitrify the sewage. However, in a number of modern activated sludge plants dealing with fairly weak sewages containing a low proportion of inhibitory trade wastes and obtaining their power cheaply from sludge gas, the aim is to achieve a considerable degree of nitrification (cf. p. 104).

PERCOLATING FILTERS

Percolating filters (also called sprinkling or trickling filters), were first introduced on a large scale in 1893 at the Salford Sewage Works, Lancashire, by Joseph Corbett. They have since been used all over the world and have taken their place as a popular, rugged and dependable means of sewage purification, especially for strong and difficult industrial sewages.

They consist of circular or rectangular beds, about 1·8 m (6 ft) deep, of well-graded media (e.g. clinker, slag, stone, gravel, coke or even coal), usually about 40 mm (1½ in) in size but increasing to 100–150 mm (4–6 in) for the bottom 0·3 m (1 ft), enclosed in walls of brick or concrete. In general, smaller media produce better effluents than larger media, but small media tend to choke more easily. The Water Pollution Research Laboratory found that, using Stevenage domestic sewage 25 mm (1 in) slag and 25 mm (1 in) clinker gave much better effluents than did similar size gravel or rock. The worst effluents were obtained from 65 mm ($2\frac{1}{2}$ in) smooth gravel. Since, however, the selection of a particular material may be decided by easy local availability or by cheapness, it is not practicable to prescribe one type or grade of media for every filter. Synthetic plastic media (synthetic resins) were first developed in the U.S.A. and are coming into use in this country. They are particularly useful for the biological treatment of trade wastes at high dosages and loadings; an advantage is that where space is restricted they can be stacked to a great height because of their lightness but a drawback is their expense, £27·5 per cubic metre (£21 per cubic yard)[3]. For sewage they might find application for partial treatment or in a 'roughing' filter. I.C.I. Ltd., have put on the market a new synthetic medium known as 'Flocor' consisting of alternating plain and shaped thin sheets of polyvinyl chloride.

A British Standard (B.S. 1438:1948) specifies the most important properties of the traditional medium; these include freedom from dust, durability, resistance to the action of sewage and trade wastes, a high surface area: weight ratio, sufficient roughness to encourage the formation of a biological film which is essential for the purification of sewage, uniformity in size (i.e. length, breadth and height nearly the same) and a shape giving a high percentage of voids so as to permit easy passage of air. The media must be placed in position carefully, for instance by hand forking or by passing over a coarse screen so as to avoid getting pieces that are too small. Care must be taken to prevent the media being crushed or compacted by excessive walking over it or by the passage over it of vehicles or trucks. A suitable underdrainage system is provided to take away the purified sewage. The whole process is continuous and, being aerobic, requires adequate ventilation of the beds. The settled sewage is sprinkled on circular beds by rotation distributors (*Figure 13*), the rotation being effected by reaction, by waterwheels or by electric motors. Travelling distributors sprinkle the sewage on rectangular filters and these are driven be electric motors or by waterwheels. Sufficient head must be available to drive the reaction type distributors. The old type of fixed distributor was inefficient, required a greater head and is now virtually obsolete. The filter distributors must, or course, be capable of dealing with $3 \times$ D.W.F. and the feed pipes should, as far as possible, allow equal distribution to all the filters. If there is continuous feed to rotating reaction type distributors, at times of low flow sewage will start to dribble from the jets and the distributor will cease to rotate. For this reason, dosing syphons are often installed so that sewage can be stored in the syphon chamber until the syphon comes into operation and supplies a flow sufficient to cause the distributor to rotate. An alternative method is to fix weirs on some

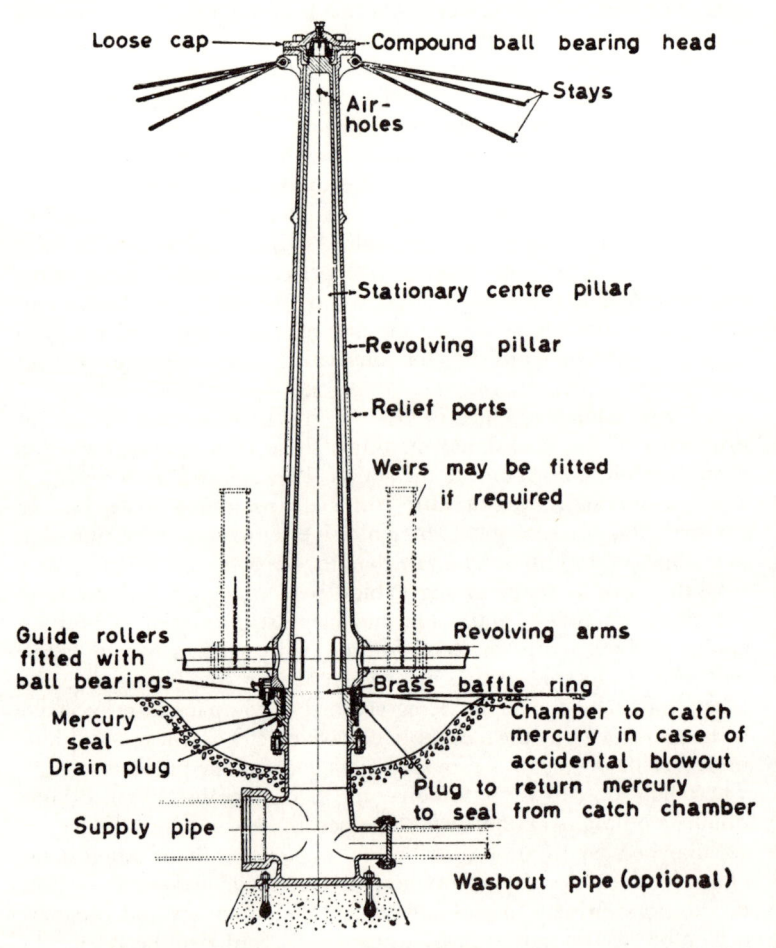

Figure 13. Simplon revolving sprinkler. (By courtesy of Messrs Ames Crosta Mills & Co., Ltd., Heywood)

of the distributor arms so that sewage discharges from the remainder only at times of low flow and has a sufficient pressure to cause the distributor to rotate.

When starting new filters, the sewage passes through the beds almost unchanged at first. Soon, however, the media become covered with an active gelatinous film containing bacteria, protozoa, fungi, etc. which, in the presence of sufficient oxygen, brings about the biological purification and stabilisation of the sewage. The time required for the maturing of a filter bed, i.e. the establishment of a well-balanced biological film, may be only a few weeks during the summer months but may take several months during the colder period of the year. It is, therefore, best to start up a filter in spring and summer.

The effluent from percolating filters contains much dark brown suspended matter ('humus sludge') washed off the filter media and requires settlement in special tanks ('humus tanks') before discharge to a watercourse.

The capacity of these tanks should be at least 4 h at D.W.F., i.e. $1\frac{1}{3}$ h at peak flows of $3 \times$D.W.F. In the case of upward flow tanks the upward flow velocity should not exceed 1·8 m per h (6 ft/h) at maximum flow. The tanks should be desludged frequently (at least once per week and preferably once per day), as humus sludge becomes septic fairly quickly and rises to the surface, especially in warm weather, thus producing unsatisfactory effluents. The earliest humus tanks were of the rectangular horizontal flow type, manually desludged. This made frequent desludging difficult, and there was a reduction in sedimentation capacity when a tank was being desludged, at least two such tanks should be provided at small works, and more at the larger works.

Modern humus tanks, except at some small works, usually take the form of one of the following types:

(a) Rectangular horizontal flow tanks with hopper bottoms
(b) Hopper-bottomed upward flow tanks
(c) Circular radial flow tanks fitted with sludge scrapers.

Sludge can be removed from these tanks as frequently as desired without taking the tanks out of operation. The hydrostatic head required to lift humus sludge is less than that needed to lift primary sludge, 0·6 m (2 ft). Humus sludge is often returned to the channel feeding the primary sedimentation tanks and settles out in these tanks with the rest of the solid matter where it tends to assist sedimentation; humus sludge is considered easier to dewater when mixed with primary sludge than on its own.

On those works where recirculation is practised, larger humus tanks have to be provided, the increase in size depending on the recirculation ratio.

The flow and strength of sewage dealt with by percolating filters should not fluctuate too greatly, otherwise the effluent may deteriorate in quality. Exceptionally strong wastes (e.g. gas liquor or kier liquor) should, therefore, as far as possible be evenly balanced out over the day. For settled sewage of average strength with a 5-day B.O.D. of, say, 220–300 mg/l, the usual volumetric rate of dosage to the filters so as to produce an effluent passing Royal Commission standards is 420 l/m^3 (70 gal/yd^3) of media per day, corresponding with a B.O.D. loading of about 0·09–0·12 kg/m^3 day (0·15–0·2 lb/yd^3 day)*. When weak sewage (B.O.D. about 150–200 mg/l) is treated, a higher volumetric loading is permissible, namely 600 l/m^3 day (100 gal/yd^3 day), equivalent to a loading of 0·12 kg/m^3 day (0·2 lb of applied B.O.D./yd^3 day)*. Strong domestic and industrial sewages require much lower volumetric loadings, viz. 210–270 l/m^3 day (35–45 gal/yd^3 day), to produce satisfactory effluents.

In double (or two-stage) biological filtration, which is not often used, two percolating filters are arranged in series and these are generally half the depth of a single filter. Two-stage filtration has some advantages in the treatment of certain strong industrial sewages, e.g. the sewage at Burton-on-Trent which contains a large proportion of brewery wastes.

Percolating filters contain large numbers of scouring organisms, such as worms, fly and insect larvae, spiders and other invertebrate animals (e.g. *achovutes viaticus*), which feed upon the biological film of bacteria, protozoa, algae and fungi. These scouring organisms play an important part in preventing excessive accumulation of this biological film on the filter media, but during the winter they tend to retreat deeper into the bed, and so there is then a tendency for the filter to become choked, which is called 'ponding' of the bed. Ponding can also be caused by overloading a filter or by dosing with sewage containing too much suspended matter. With the arrival of warmer weather in spring and early summer, the scouring organisms return to the surface, feed on the surface growths and thus relieve ponding. At this time, excessive amounts of suspended solids are discharged with the effluent owing to solids becoming loosened, and the filter is said to be 'sloughing' or 'unloading'. Several remedies are available for overcoming excessive ponding, for example forking the surface of the bed, flushing with plant effluent and chlorination of the sewage. Fine-mesh screens installed in the feed channels to the filters have in many cases helped to prevent ponding. Excessive film growth on circular filters can be controlled by slowing down the distributor speed by means of a variable speed drive (see Chapter 12). The best

---

* kg of B.O.D. in y m^3 of sewage = $\dfrac{y \times \text{B.O.D. of sewage (mg/l)}}{10^3}$; the B.O.D. is that of the *settled* sewage.

way of controlling film growths, however, is by the adoption either of alternating double filtration or of recirculation.

The operation of filters is sometimes associated with the production of unpleasant smells, the extent of the nuisance depending on the character of the sewage, the temperature and atmospheric conditions. Filters should, therefore, be located well away from houses.

A nuisance commonly associated with the operation of the conventional filter is the presence of large numbers of small flies or midges, such as *Psychoda alternata, Psychoda severini* and *Anisopus fenestralis*. These are objectionable not only to workers at the sewage plant but also to people living in the vicinity. Among the remedies suggested for dealing with the fly nuisance are flooding the filter once a week for 24 h and the use of chlorine and other chemicals. An effective modern remedy is the application to the filter of intermittent doses of a synthetic insecticide such as DDT or Gammexane, which kills the adult flies without interfering with the bacterial flora or with the purification of the sewage.

Although the percolating filter has a considerable reserve capacity, any marked overloading will lead to ponding or to production of inferior effluents. Nevertheless, it is possible to increase the capacity of a filter without impairing effluent quality. The following methods, usually referred to as 'high-rate filtration' (though the filters are not 'high-rate' in the American sense), have been developed in recent years.

Alternating double filtration—This process[4] depends upon an observation originally made by Whitehead and O'Shaughnessy of Birmingham that a ponded or clogged filter could be restored to a healthy condition by dosing it with partially purified effluent from another filter. In this way, the biological film and growths causing the ponding are disintegrated and removed. The method is applied by arranging two filters in series, dosing the primary filter with settled sewage and pumping the settled effluent from this to a secondary filter. After an interval varying from a day to a week, or when the primary filter shows marked signs of ponding, the sequence is reversed, the primary filter being used as secondary filter and the former secondary becoming the primary. In this way, at least twice the normal volume of settled sewage can be treated to Royal Commission standards on the same total cubic capacity of filter media. The process is well adapted to relieving overloaded filters at comparatively small cost, since it is only necessary to install extra humus tanks and to provide extra pumping facilities. This is providing that the feed pipes and distributors are capable of dealing with the additional volumetric load.

At Salford, Lancashire, where there is a highly industrial sewage which causes little or no ponding on the filters, it has been shown that alternating double filtration has no advantage over single filtration.

Enclosed aerated filters—Another high-rate process, used to some extent in Germany and South Africa, is the enclosed aerated filter[5]. This is an enclosed filter, 3·6–5·5 m (12–18 ft) deep, provided with an air-tight cover and ventilated by means of a current of air introduced *downwards* through the bed by mechanically driven fans. In this way, over twice as much sewage can be treated as in the conventional open 1·8 m (6 ft) filter. Even more remarkable results have been reported by Tedeschi and Lucas[6] with a new type of enclosed filter in which a natural draught of air is induced *upwards* through the medium. Results obtained by feeding sewage containing gas liquor on an open filter at 240 l/m^3 day (40 gal/yd^3 day) were comparable with those achieved by dosing this enclosed filter at 2 970 l/m^3 day (500 gal/yd^3 day). This means that the enclosed filter was dealing satisfactorily with more than 12 times as much sewage as the open filter.

Recirculation—A popular method of increasing the capacity of a filter is recirculation. This usually involves the biological filtration of settled sewage which has been diluted with a proportion of settled purified effluent (humus tank effluent) from the filters. Pumps are needed for recirculation. It has been shown by Lumb and Eastwood[7] that the effluent recirculated can be purified effluent from the same or other filter beds, or even from an activated sludge plant*, in which case pumping may not be required. The recirculation ratio, or ratio of volume of recirculated filter effluent to volume of settled sewage, is generally 1:1 but may be higher, such as 2:1. The adoption of recirculation makes it possible to dose a filter with about 2–3 times the normal quantity of primary sedimentation tank effluent, irrespective of the volume of recirculated liquor, and still produce an effluent of chemical quality comparable to that obtained by orthodox single filtration. Nitrification is, however, much less when recirculation is used. Lumb suggested that this is probably due to the zone of clarification and carbonaceous oxidation (which in an ordinary filter is confined to the top layer) extending much further down when recirculation is adopted.

Experiments with certain industrial sewages (e.g. Salford and Macclesfield) have shown that recirculation offers no advantages over single filtration. The process must not, therefore, be regarded as a universal panacea.

In those cases where the character of the sewage does not demand the provision of recirculation, it is not good practice, when designing a completely new works, to provide a recirculation system to treat

---

* Thus, the presence of nitrate is not necessary.

Aerobic Biological Treatment

the existing and immediate future flow of sewage. It is advisable to design the works to operate in the first instance on single filtration, but with provisions for recirculation to make extra treatment capacity available to deal with any future increases in volume and load.

Alternating double filtration and more particularly recirculation are being increasingly used in the British Isles in sewage schemes to relieve existing overloaded plants and to treat strong industrial sewages. The loadings are of the order 0·18–0·24 kg of applied B.O.D.*/m^3 of media daily (0·3–0·4 lb/yd^3 day). Somewhat larger size media than usual are generally favoured for these processes (e.g. 65 mm or 2½ in).

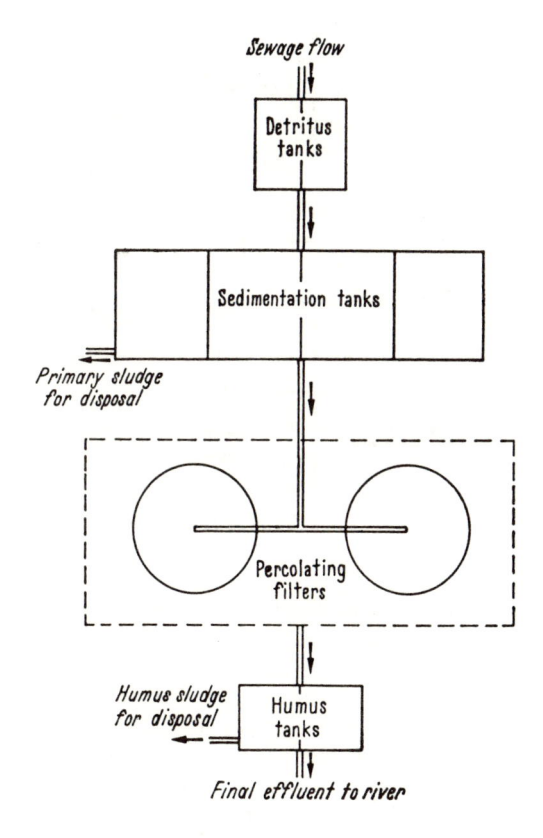

Figure 14. Flow diagram of conventional low-rate percolating filter system of sewage purification

* The B.O.D. is that of the *settled* sewage.

In the U.S.A., where sewages are usually weaker, much higher rates of filtration have been reported, loadings over the range 0·6–1·5 kg of B.O.D.*/m^3 day (1–2·5 lb/yd^3 day) being used at many works. At such high loadings, the removal of B.O.D. may be 50 per cent or less, but a sub-standard effluent may be acceptable where ample dilution is provided in the river.

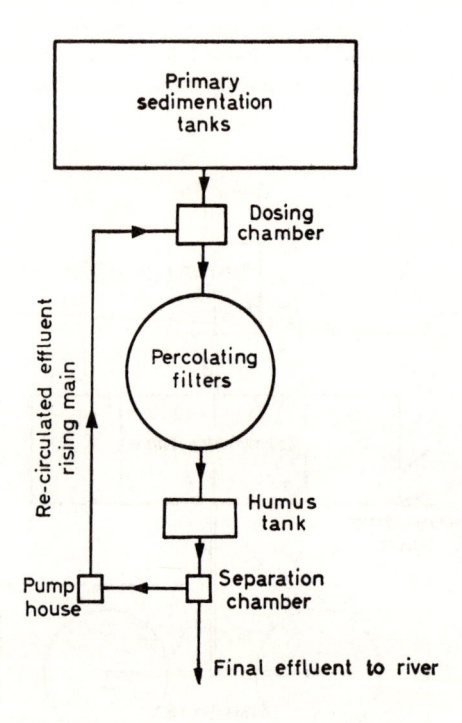

Figure 15. Flow diagram of percolating filter system of sewage purification: filtration with recirculation

High rate filters have other advantages besides increasing filter capacity. They operate without ponding and are much less subject to odour or filter-fly nuisance than are low-rate filters.

Efficiently operated low-rate percolating filters normally yield well-nitrified sewage effluents containing much nitrate but very little nitrite. Less nitrate, however, is produced in high-rate filtration processes.

* The B.O.D. is that of the settled sewage.

Aerobic Biological Treatment

The nitrifying bacteria bringing about nitrification function best at about 20–25°C, hence less nitrate is formed in winter than in summer. They require carbon dioxide to synthesise their protoplasm and the presence of sufficient bases (e.g. sodium, calcium) to neutralise the nitrous and nitric acids produced by oxidation of ammonia, free oxygen and nutrient materials, especially phosphates, magnesium and iron. The presence of organic matter is not necessary, indeed some organic compounds present in certain trade wastes tend to interfere with nitrification. Many heavy metals inhibit nitrification, e.g. nickel (10 mg/l), mercury (2 mg/l) and silver (0·3 mg/l).

The reduction of nitrates ('denitrification') can be brought about by nitrate-reducing bacteria when little or no dissolved oxygen is present, for instance in percolating filters that have become ponded. Under these conditions, nitrate can be reduced to nitrite, ammonia and even nitrogen gas. The rising of sludge which sometimes occurs in humus tanks is generally attributed to the formation of nitrogen bubbles produced by denitrifying bacteria.

A flow diagram of the conventional low-rate filter is shown in *Figure 14* and a diagram of a recirculation scheme in *Figure 15*.

ACTIVATED SLUDGE PROCESS

In 1913, Arden and Lockett[2], as a result of experiments on the aeration of Manchester sewage, announced the discovery of a new and rather revolutionary method of sewage treatment—the activated sludge process. This consists essentially in the aeration (for a sufficient length of time) of a mixture of settled sewage with a special bacteriologically active sludge; after settlement, an effluent is obtained of a quality similar to that of a good percolating filter effluent. In the original laboratory experiments, it was found that by aerating sewage for several weeks, oxidation and nitrification occurred and a dark brown flocculent sludge was deposited. After settlement, the purified sewage was withdrawn and the sludge used to assist the oxidation of a further quantity of sewage. This was repeated many times, the accumulated sludge being always retained, when it was found that purification was occurring with progressively shorter aeration periods as the sludge became more active. Finally, a highly active sludge was obtained (termed 'activated sludge') which could achieve purification of sewage in a few hours.

On the large scale, it is now possible to build up a supply of activated sludge in several weeks by aeration of sewage with return of the sludge. The utilisation of sludge from another activated sludge plant, or even inoculation with humus sludge, can greatly reduce this time.

The activated sludge process is now used at cities and towns in many countries, being particularly attractive for large cities where land is scarce, since an activated sludge plant occupies only a fraction of the area of a percolating filter installation.

It is of particular importance to ensure that the sludge in the mixture of sewage and activated sludge be kept in suspension by sufficient turbulence and that adequate amounts of oxygen are present. This can be done on a large scale by the use of diffused air or by the adoption of one of the many mechanical aeration systems.

DIFFUSED AIR SYSTEM[8]

This was the original method of introducing air (in the form of fine air bubbles) used by Ardern and Lockett. Filtered compressed air (0.006–0.009 m^3/l or 1–$1\frac{1}{2}$ ft^3 of air/gal of sewage treated) is bubbled by means of porous tiles or domes made of alundum (fused aluminium oxide) through the mixture of sewage and activated sludge (so-called 'mixed liquor') in aeration tanks commonly 2.4–3.7 m (8–12 ft) deep. A few of the older aeration tanks are as much as 4.5 m (15 ft) deep. The amount of activated sludge in this mixed liquor is usually maintained between 1 500 and 3 000 mg/l of suspended solids.

The air diffusers were formerly arranged at the tank bottom in (a) the ridge and furrow system, in which the diffusers are set transversely or longitudinally at the bottom of depressions between ridged spaces, so that there are no surfaces on which sludge can settle, or (b) the spiral flow system in which the diffusers are placed near one side of the lower part of the tank wall, so as to induce a spiral turbulent flow.

In new plants, the more modern Dome diffusers are arranged along the bottom of a flat floor.

The detention period in the aeration tanks depends upon the strength and character of the sewage; in Britain it ranges from about 6 h for weak sewage to about 12 h for strong sewage at D.W.F. Expressed in terms of the applied B.O.D. of settled sewage, the normal loading in these plants is roughly 480–800 kg of 5-day B.O.D. /1 000 m^3 of aeration tank capacity per day (30–50 lb/1 000 ft^3 day) or, say, around 0.6 kg of B.O.D./m^3 of tank daily (1 lb/yd^3 day). In the U.S.A., aeration periods are usually less (4–8 h) because the sewages are weaker.

The diffused air system, a proprietary method of Messrs. Activated Sludge Ltd., London, is used at a number of important works,

Aerobic Biological Treatment

e.g. Mogden (Greater London Council), West Hertfordshire (Colne Valley), St. Helens and Hogsmill Valley (Greater London Council).

The diffused air system suffers more from foaming troubles due to synthetic detergents in sewage than the various mechanical aeration processes. Although foaming can be controlled by the addition of certain anti-foam chemicals, the cheapest method is to spray the foam with final effluent.

The older flat-plate diffusers tended to clog rather easily, especially in the presence of iron salts, but this tendency is considerably lessened by the use of the newer Dome diffusers[9]. All types of diffuser require cleaning every few years to remove dust, organic materials, etc. This can be done by burning the diffuser or by treatment with chemical reagents such as hydrochloric acid, nitric acid or caustic soda.

A new diffused air process using coarse air bubbles (the INKA process) is described in Chapter 12.

MECHANICAL AERATION SYSTEMS

The rate of oxygen transfer by various mechanical aeration devices is affected by the nature of the aeration device, depth of submergence, temperature, turbulence in the tank, depth of the tank and the chemical character of the waste. Improved results in recent years have been achieved by modifying the ways in which more turbulence is produced. This increases the rate of oxygen input into the mixed liquor, or the so-called 'oxygenation capacity', i.e. the rate of oxygenation expressed as g. of oxygen per hour per cubic metre of aeration tank volume, using deoxygenated distilled water at 10°C and 101 000 N/m^2 (760 mm) barometric pressure[10].

1. *'Simplex' surface aeration*[8]—In these plants, made by Ames Crosta Mills & Co. Ltd., Heywood, Lancashire, an inverted, rapidly rotating cone provided with steel blades draws the sewage up a steel pipe in a hopper-bottom aeration tank (about 4·4 m or 14½ ft total depth) and sprays it across the liquid surface, thus causing intense aeration of the mixed liquor. The aeration period, originally about 15 h, has been reduced to about 8–9 h by the development of a later type of 'high-intensity' aeration cone which differs from the older type in proportions and in the form and disposition of the blades (*Figure 16*); at least 50 per cent more sewage can be treated than with the older type. Work at Manchester by McNicholas and Tench[11] demonstrated that still further improvement can be effected by reducing the depth of tank to 1·8 m (6 ft), by using a newer design of high-intensity cone and by operating at a somewhat higher cone speed. In

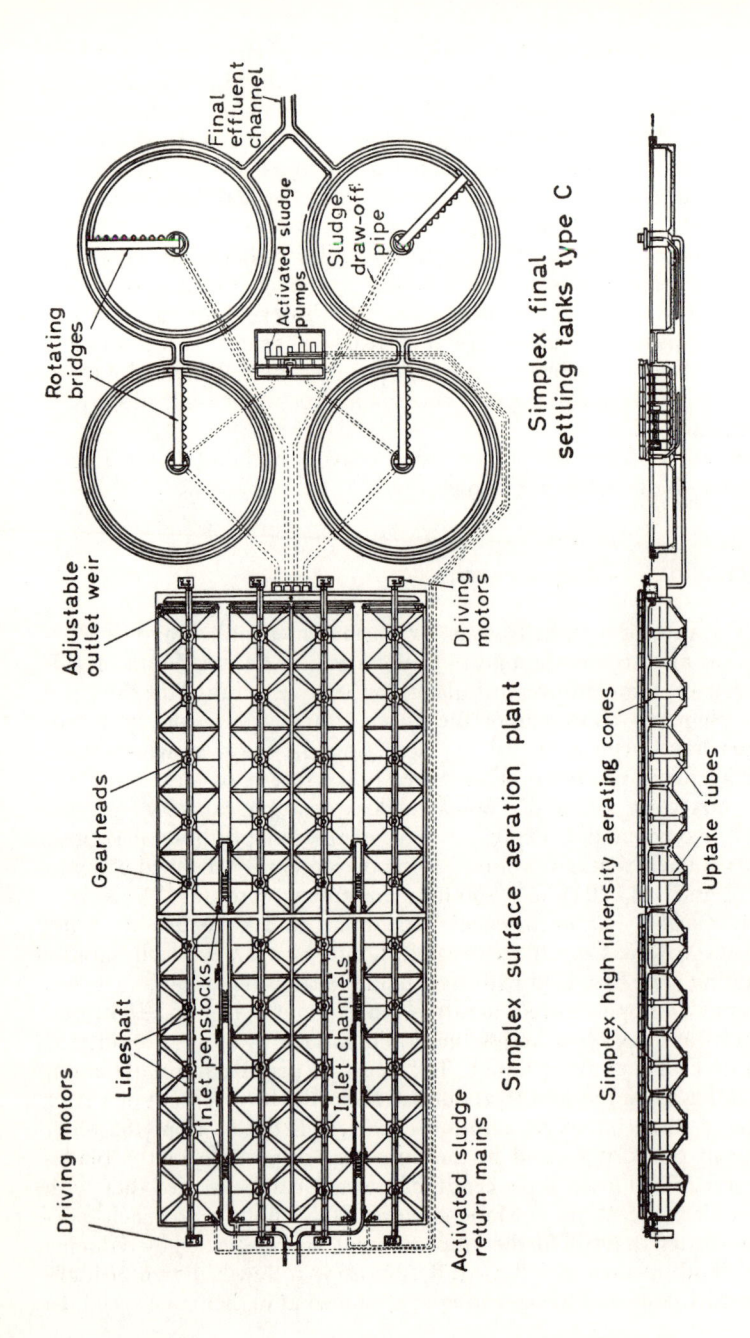

Figure 16. Simplex high-intensity cone plant. (By courtesy of Messrs Ames Crosta Mills & Co., Ltd., Heywood)

Aerobic Biological Treatment

this way, the aeration period required to produce a Royal Commission effluent was reduced to just over 2 h, with very little increase in power consumption. The concentration of activated sludge in the mixed liquor has to be rather greater (3 000–5 000 mg/l suspended solids) than in the conventional type of plant and is adjusted so as to maintain a dissolved oxygen concentration of 2–3 mg/l.

Many of the largest sewage works are now using the Simplex system, outstanding examples being Davyhulme (Manchester) and Crossness (Greater London Council).

Another mechanical device giving surface aeration similar, in some respects, to that of the Simplex cone is the Simcar aerator.

2. *Kessener Brush System*—This process was introduced in Holland by Dr. H. H. Kessener in 1925. The sole licensees in Great Britain are Messrs. Whitehead and Poole Ltd., Radcliffe near Manchester. Circulation and aeration are achieved by rapidly (over 100 rev/min) revolving stainless steel brushes which are partly submerged in the mixed liquor along one side of the aeration tank and produce a spray of fine droplets as well as waves along the surface; more intense aeration can be obtained by increasing the speed or immersion of the brushes. A number of sewage plants on the continent use this system, and in Great Britain, Huntingdon and Godmanchester, Hawick (Scotland), Stockport and Rochdale. Several trade waste purification plants also operate on this system.

Work is proceeding on the development and improvement of the Kessener Brush so as to get even more intense aeration. Pasveer[12] in Holland has shown that by reducing the depth of the aeration tank from 4·0 to 1·8 m (13 ft to 6 ft), a very much more rapid purification of sewage can be achieved; a Royal Commission effluent can be obtained in this way from a domestic sewage by aeration for only 1 h. It was also claimed that some benefit was obtained by covering each brush with a hood. Pasveer considers that local foaming around the brush causes most of the oxygenation that occurs. McNicholas and Tench[11], who also carried out experiments along these lines with the difficult Manchester industrial sewage, using an aeration period of about $1\frac{1}{2}$ h in a pilot plant, produced an effluent just above the Royal Commission B.O.D. standard (28 mg/l instead of 20 mg/l). A typical plant layout is shown in *Figure 17*.

Newer designs of high intensity rotors enabled Todd[13] at Rochdale to obtain effluents of Royal Commission standard with low aeration periods (1·23–1·69 h) and with a power consumption of only 4·4 kW/10^6 l (27 hp/10^6 gal).

The Kessener Brush, or similar rotor as a means of aeration, has been applied by Pasveer to an inexpensive method of dealing with wastes from small communities (see Oxidation Ditch, Chapter 11).

3. *Sheffield Bio-aeration System*—In this system, used at Sheffield, Stockport, Wolverhampton, and Stalybridge, the sewage passes along very long, shallow channels, about 1·2 m (4 ft) deep, in which circulation is brought about by vertical revolving paddles which set up a wave motion inducing aeration. The period of aeration is about 15 h. Edmondson[8] greatly improved this process by substituting triangular paddle blades for the older rectangular ones and by increasing the speed of rotation from 15 to 30 rev/min. These modifications allowed about 60 per cent more sewage to be treated at Sheffield. When the old bio-aeration plant at Manchester (Davyhulme) was altered in this way, about 40 per cent more sewage was

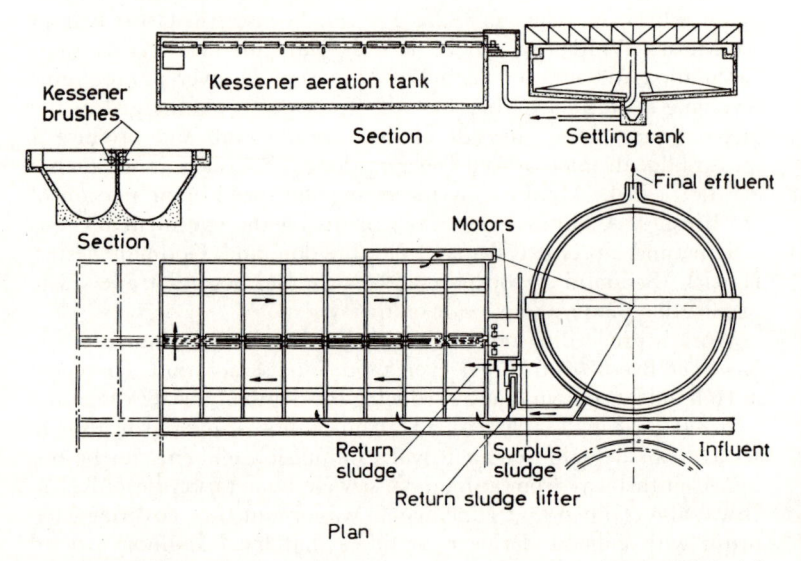

Figure 17. Typical plant lay-out: Kessener brush system. (By courtesy of Messrs Whitehead & Poole Ltd., Radcliffe)

dealt with; there was also a 44 per cent reduction in power consumption and a reduction in detention period from 17 to 12 h.

At Chesterfield, Hirst[14] obtained improved operation of an experimental bio-aeration unit by using four lines of paddles instead of the usual two lines. This increased treatment capacity by 80 per cent and reduced the aeration period by about 45 per cent. Extensions to the plant will be undertaken along these lines.

The activated sludge process is, like biological filtration, a continuous process. The various steps can be summarised as follows:
  (i) The settled sewage passing to the inlet of the aeration tanks is

mixed with sufficient return activated sludge to give a suspended solids concentration of 1 000–5 000 mg/l, or about 10–25 per cent by volume as measured after 1 h settlement in a graduated litre cylinder. The optimum concentration of activated sludge must be determined experimentally for each plant, since it varies with local conditions, such as the type of plant, the tank capacity, the character and strength of the sewage, the air supply, and the quality of effluent required.

(ii) The resulting mixed liquor is aerated for a suitable period in the aeration tanks. It was thought at one time that the concentration of dissolved oxygen should not fall below 1–2 mg/l, but recent work seems to show that, for a non-nitrified effluent, a concentration of 0·5 mg/l in the mixed liquor is sufficient[14a].

(iii) At the outlet end of the aeration tanks, the mixed liquor is allowed to settle in final settlement tanks for about 2–3 h (the time tends to vary depending on the design of the tank). The supernatant liquid, constituting the purified effluent, is discharged to a watercourse.

(iv) Part of the activated sludge is then returned to the aeration tank inlet and mixed with the settled sewage, as under (i). In some instances, particularly with diffused-air plants or where strong sewages are being treated, the activated sludge is conditioned by re-aeration in special tanks before return to the aeration tank inlet. Usually one-fifth to one-tenth of the total number of aeration tanks are provided for this purpose. When 'step aeration' is used re-aeration is not necessary. The amount of activated sludge returned to maintain the process, expressed as a percentage of the sewage D.W.F., is normally about 50–70 per cent, but may be as low as 25 per cent or even as high as 100 per cent.

(v) The surplus activated sludge not used in (i) is disposed of in various ways (see Chapter 8).

Sludge must be removed continuously from the final settlement tanks so that it can be returned in as fresh a condition as possible. These tanks are usually hopper-bottomed upward flow tanks or circular radial flow tanks fitted with mechanical sludge scrapers. A smaller hydrostatic head [0·6 m (2 ft)] is required to lift activated than primary sludge. Surplus activated sludge is often returned to the channel feeding the primary sedimentation tanks and settles out in these tanks with the rest of the solid matter. Like humus sludge, it tends to assist sedimentation and is easier to dewater when mixed with primary sludge than on its own.

A flow diagram of the activated sludge process is shown in *Figure 18*.

To secure economy, or better utilisation, of air, modified forms of aeration have been tried at certain plants in the U.S.A., said to give better or more flexible control and to reduce operation costs. Some of these will be briefly described.

Tapered aeration[15]—Air is one of the largest operating costs in the activated sludge process. Now, the oxygen requirements of the process are greatest at the inlet end of the aeration tanks where the

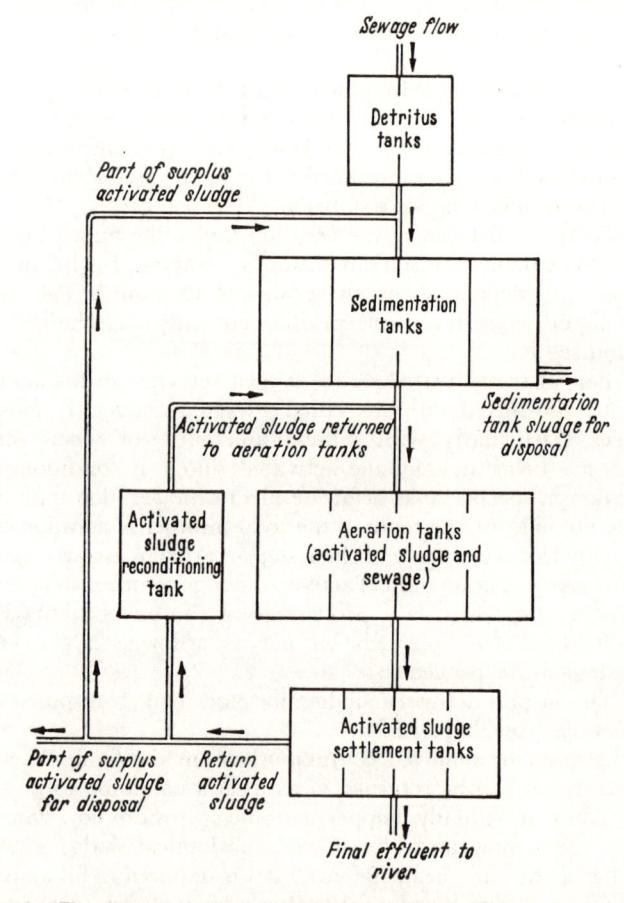

Figure 18. Flow diagram: Activated sludge process of sewage purification

full sewage load enters, and least at the outlet end. To meet these requirements and to secure more economical operation it has been proposed that the air supply should be increased at the earlier stages (e.g. by increasing the diffuser area at the inlet or by using larger amounts of air there) and decreased during the later stages, a modification termed 'tapered aeration'.

Step aeration[16]—Another method of securing greater economy in operation and better control is 'step aeration', introduced by Gould in the U.S.A. In the conventional method of operation, the entire sewage load is introduced with the return activated sludge at the inlet end of the aeration tanks. With strong sewages this can result in the dissolved oxygen content of the mixed liquor falling to zero. In step aeration, this is obviated by adding a portion (say $\frac{1}{4}$) of the sewage load at, say, each of four inlets. The return activated sludge enters at the beginning of the aeration tank and so does not receive the full sewage load as it would in the normal method. It is evident that step aeration involves the use of higher concentrations of suspended solids in the mixed liquor in the earlier stages but obviates any need for re-aeration of the return sludge in a special tank.

High-rate activated sludge process—The work of Gould and of Setter[17] in the U.S.A. has led to the introduction of a high-rate process involving the use of much less activated sludge in the mixed liquor and much shorter aeration periods. For instance, at the New York (Jamaica) plant, only 650 mg/l of activated sludge were carried in the aeration tanks (instead of the usual 2 000 mg/l) and the detention period was reduced to 2 h. The sludge age is low, only one-tenth of the normal age. The effluent is not of Royal Commission standard and the B.O.D. is reduced by only about 75 per cent. The process might find application in the British Isles at estuarial or coastal towns where a degree of purity intermediate between a sedimentation tank effluent and a Royal Commission effluent may be required. The B.O.D. loading is about three times the normal loading, i.e. about 1·6 kg applied B.O.D./m³ of aeration tank daily (2·7 lb/yd³ day).

Another high-rate process, used at Wuppertal (Germany)[18], differs from the conventional process in using aeration tanks about one-third the usual size. The B.O.D. reduction is approximately 70 per cent, with an aeration period of only 1·75 h, and the B.O.D. loading is nearly four times the normal figure. A sludge age of about 1–2 days gave the most favourable results.

The activated sludge obtained in these high-rate processes is dense, settles quickly and does not 'bulk' as easily as normal activated sludge. The cost is much less than that of the conventional process.

Extended aeration[19, 19a, 20]—This process, also termed 'total oxidation' or 'aerobic digestion', originated in the U.S.A., where it was found that by the use of prolonged aeration periods (24 h or larger) and a high rate of return sludge (up to three times the volume of incoming sewage), it is possible to bring about considerable oxidation and aerobic digestion of the activated sludge solids. The rate of accumulation of the sludge is, therefore, smaller than in conventional plants, thus making the process particularly suitable for small plants. There are now in this country several self-contained prefabricated or 'package' sewage plants embodying the 'extended aeration' principle.

The aeration period should be at least 24 h and preferably higher, e.g. 2 or 3 days. A high concentration of activated sludge is maintained in the mixed liquor, often as much as 4 000–8 000 mg/l. The loadings are of the order 0·25 kg or less of applied B.O.D. per m^3 of aeration tank capacity per day (15 lb/1 000 ft^3 day) (i.e. much lower than in conventional plants). Well-nitrified effluents of Royal Commission standard can be produced. Since high contents of suspended solids are sometimes found in the effluents, it is necessary to remove sludge from time to time and also to provide some means of 'polishing' the final effluents, e.g. on grass plots or sand filters. If the plant is working efficiently the waste sludge removed is so well-oxidised that it is comparatively easy to dry on sand beds. Foaming in these plants can be prevented by seeding with sludge from another plant or with humus sludges.

GROWTH OF ACTIVATED SLUDGE ORGANISMS

When activated sludge purifies sewage in the presence of dissolved oxygen, the bacterial cells grow in size until each cell divides into two daughter cells. The process continues until:

(a) the food supply is exhausted (i.e. when the sewage is completely purified), or

(b) other factors limit the growth, e.g. lack of nutrients, deficiency of dissolved oxygen, and build-up of toxic end-products.

Normally, something like 60 per cent of the organic matter in sewage, as measured by the 5-day B.O.D. test, is assimilated to form new sludge cells.

A typical growth curve of activated sludge organisms is shown in *Figure 19*.

Hopwood and Downing[21] express the rate of growth of sludge in terms of a 'sludge growth index', defined as the weight of activated sludge formed per unit weight of 5-day B.O.D. applied. This index is about 0·55 for a conventional plant treating domestic sewage, and about 0·9 for a high-rate plant. The index decreases with increasing sludge concentration and with increasing aeration period, but only increases to a small extent with stronger sewages.

The growth curve can be divided into the following phases[22]:

(i) *Lag phase*—At the beginning, there is sometimes a lag phase of slow growth of cells. This does not normally occur in the treatment of sewage by activated sludge as the organisms would be in an active

Aerobic Biological Treatment

state. It only occurs when adaptation of the bacteria to a new waste is taking place and so may happen with many trade wastes unless acclimatised seed material is used.

(ii) *Logarithmic growth phase (or log phase)*—This is a phase of increasing growth to constant growth. A more rapid rate of cell growth occurs here and the rate of cell multiplication is at a maximum because the

$$\frac{\text{Food}}{\text{Cell}} \text{ ratio} = \left(\frac{\text{kg (lb) of B.O.D.}}{\text{kg (lb) of aeration solids}}\right) \text{ is high.}$$

In the conventional activated sludge process, the maximum rate of removal of B.O.D. takes place at this stage. Moreover, since the

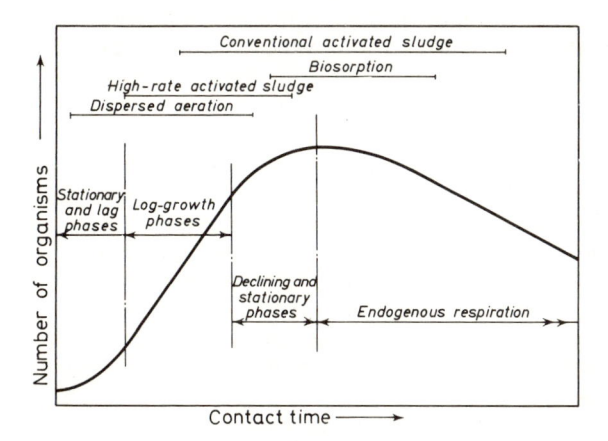

Figure 19. Typical growth curve for activated sludge. (Reproduced from P.C.G. Isaac, Waste Treatment, by courtesy of Pergamon Press Ltd.)

sludge increases in quantity on account of the very rapid growth of micro-organisms, surplus sludge *not* required in the process must be regularly removed.

(iii) *Declining growth phase*—Growth of cells starts to decline here as the food supply diminishes with the breakdown and purification of the sewage. The growth curve reaches its maximum here. The food: cell ratio is low and is decreasing. The rate of removal of B.O.D. decreases as the organic matter is consumed.

(iv) *Endogenous respiration (or maximum stationary) phase*—The growth curve falls rapidly here as the organic matter becomes exhausted. The insufficient food supply compels the organisms to utilise organic matter *within* the cell. This is so-called 'endogenous respiration' since dissolved oxygen is still required for this phase. It is really a process of auto-digestion or 'aerobic digestion'. It only occurs to a limited extent in the conventional activated sludge process. But, it predominates in, and is a characteristic feature of, the extended aeration process.

After this phase, growth ceases owing to the death of the organisms due to exhaustion of nutrients and to production of metabolic poisons.

'QUALITY' OR 'CONDITION' TESTS OF ACTIVATED SLUDGE[23]

The quality of an activated sludge is an important characteristic upon which its ability to efficiently purify sewage depends. A good quality sludge is usually golden brown in colour and has a pleasant 'earthy' smell, provided it is kept well aerated. If deprived of oxygen for any length of time, however, it darkens and eventually becomes black, septic and foul. With sewage plants on the combined sewerage system, the quality of the activated sludge improves after a spell of wet weather and deteriorates during a long dry spell. The better results obtained in wet weather may be due to dilution of the sewage by well oxygenated rain water combined with the beneficial effect of soil protozoa washed in by the rain. Indeed, workers at the Water Pollution Research Laboratory[23a] have proved that ciliated protozoa are essential for the conventional activated sludge process*. The 'quality' or 'condition' of a sludge can be diagnosed on the basis of a number of tests, of which the more important will be briefly given.

Sludge activity—Sludge activity can be determined by measuring the rate of oxygen utilisation of sewage-sludge mixtures either in an apparatus termed an 'Odeeometer' in which the 'Nordell number'[24], i.e. the amount of oxygen (in mg/l) utilised per hour by the sludge, is determined, or alternatively, manometrically in a micro-respirometer such as the Warburg apparatus.

Microscopic examination—Ardern and Lockett[23] showed that the microscopic examination of activated sludge gave very valuable indications of its quality*.

Satisfactory sludges have a preponderance of ciliate protozoa which move and feed with the aid of cilia, or fine hair-like processes. Among these ciliates are *Carchesium, Vorticella, Euplotes, Epistylis, Loxophyllum, Choenia, Chilodon, Colpoda, Colpidium, Paramecium* and *Aspidisca* (see *Figure 20*). Very few flagellate protozoa and very few

* Diagrams of ciliated protozoa in activated sludge are given in a Ministry of Technology publication, *Technical Paper No. 12*, 1969. London: H.M.S.O.

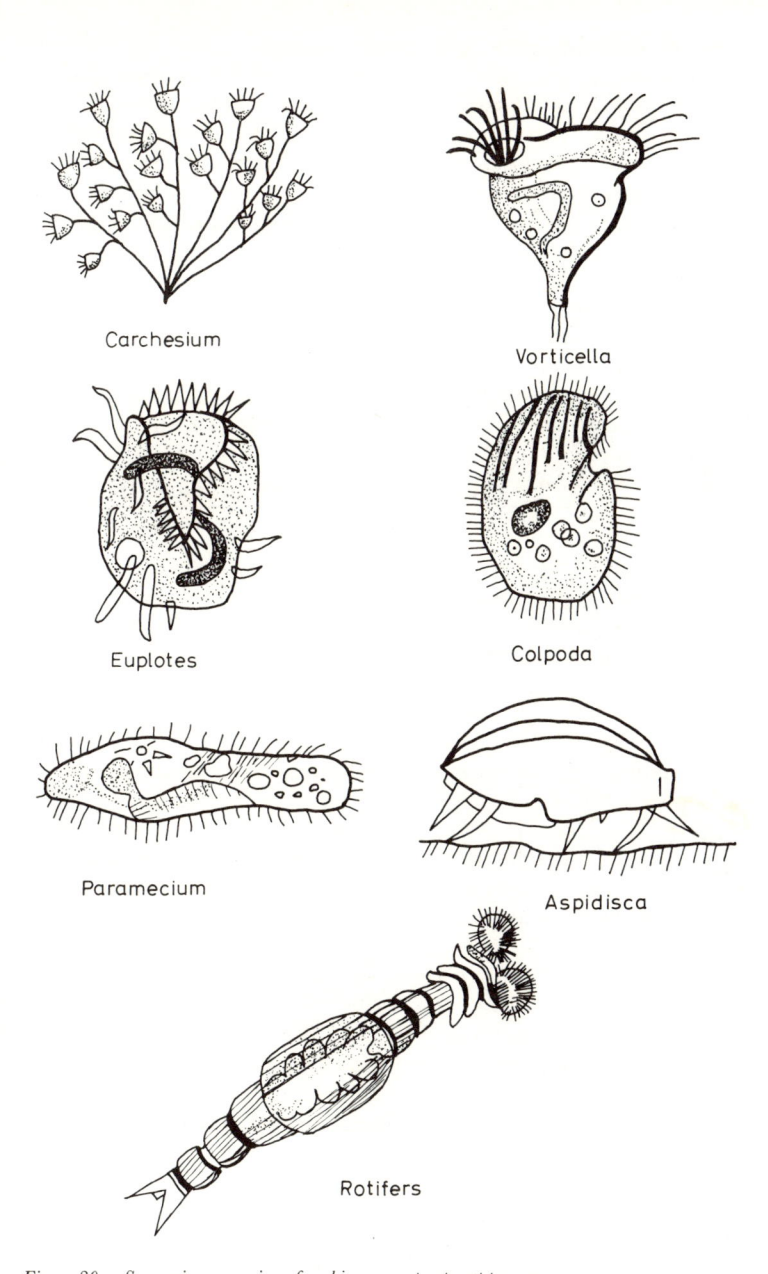

Figure 20. Some micro-organisms found in sewage treatment processes

amoebae and rhizopods are present in good sludges. Filamentous growths are absent, or present in only small amounts. Rotifers are often present. Unsatisfactory and bad sludges contain predominantly flagellate protozoa which have one or more flagella or fine whip-like processes, while very few ciliates are present. Filamentous growths are usually present in bad sludges. These observations have been confirmed by later workers but apply only to sludges obtained in the conventional method of operation. In high-rate processes, the sludge contains bacteria but no protozoa.

Sludge index[25]—Among the most frequently used sludge indexes for plant control are the Mohlman sludge volume index (S.V.I.) and the Donaldson sludge density index (S.D.I.).

The sludge volume index is the volume (in ml) occupied by 1 g of sludge after settling the mixed liquor for 30 min in a litre graduated cylinder

$$\text{S.V.I.} = \frac{\text{settled volume of sludge (per cent) after 30 min}}{\text{suspended solids (per cent)}}$$

This index increases as the sludge tends to settle badly or 'bulk', and generally varies from around 40 for a good sludge to about 200 or more for a poor sludge.

The sludge density index is given by the expression

$$\text{S.D.I.} = \frac{\text{suspended solids (per cent)} \times 100}{\text{settled volume of sludge (per cent) after 30 min}}$$

This index varies from about 2·0 for a good to about 0·3 or less for a poor sludge.

Sludge age[16]—The sludge age, as defined by Gould, is the ratio of the weight of suspended solids in the mixed liquor to the weight of suspended solids introduced daily into the aeration tanks. Hence we have the expression

$$\text{Sludge age} = \frac{V \times S_M}{Q \times S_S} \text{ days}$$

where V = aearation tank volume 10^6 l (gal)
  S_M = suspended solids in mixed liquor (mg/l)
  Q = flow of sewage in 10^6 l/day (gal/day)
and S_S = suspended solids in sewage or tank effluent (mg/l).

The sludge age can be regarded as the average detention time of a sludge floc in the activated sludge plant. In conventional operation, a sludge age of 3–4 days is necessary; for the high-rate process, however, it should be about 0·2–0·4 day. Between a sludge age of 0·5 day and 3 days, sludge settles poorly, hence ages between these values must be avoided.

Sludge settleability—Activated sludge in good condition settles very rapidly from mixed liquor whilst poor-quality sludge settles slowly. This is illustrated by typical curves showing the rate of settlement of good, fair and poor sludges (*Figure 21*).

When activated sludge plants are overloaded (e.g. with shock loads of highly carbonaceous waste), the sludge loses its property of rapid settlement and settles badly in the final settlement tanks, so that it tends to be carried over and to contaminate the final effluent. This troublesome phenomenon, termed 'bulking', is generally confined to the smaller plant where, perhaps, there may be considerable fluctuations in organic load and often insufficient technical supervision. No one remedy is effective in correcting bulking in every case, but measures such as increasing the air supply and the aeration

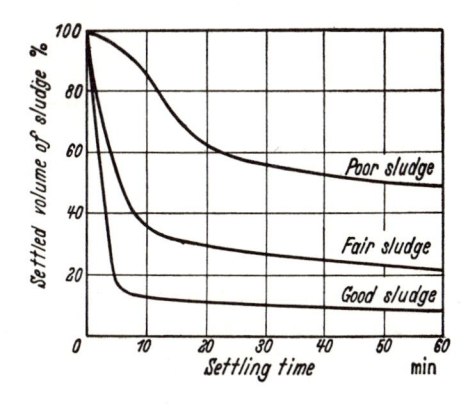

Figure 21. Typical settlement curves for activated sludge

period, addition of finely divided inert materials, reducing the suspended solids concentration in the mixed liquor, controlled chlorination and reducing the organic load by recirculation with plant effluent have been successful in some instances. In extreme cases, the bulking sludge must be disposed of (in lagoons but *not* to the river) and a new sludge has to be built up. Tests such as microscopic examination, sludge index and sludge settleability, estimation of dissolved oxygen in the mixed liquor at various parts of the aeration tanks should, if possible, be done daily. Any appearance of filamentous organisms, increase in sludge volume index or decrease in sludge density index, or marked fall in the dissolved oxygen content of the mixed liquor will act as warning signs of the development of an unsatisfactory sludge, so that control measures may be taken to correct bulking in time.

Laboratory studies by Hattingh[26] suggest that sludge bulking is

Table 14. COMPARISON OF PERCOLATING FILTERS AND THE ACTIVATED SLUDGE PROCE

Factor	Percolating filters	
	Advantages	*Disadvantages*
Initial cost	—	High
Operating costs	Low	—
Influence of weather (combined sewerage system)	Work well in summer	In winter, liable to 'ponding' difficulties and freezing
Technical control	Little needed	—
Nature of sewage	Suitable for strong and difficult industrial sewages	—
Pumping of sewage	—	Sometimes necessary since filters need a high head
Area of land required	—	Large
Fly and odour nuisance	—	Considerable
Final effluent	Highly nitrified, even if trade wastes present	Suspended solids apt to be high
Secondary sludge produced	Small in quantity	—
Synthetic detergents	Little foam on plant	—

due to the combined effect of high B.O.D.:N and high B.O.D.:P ratios in the wastewater undergoing purification. No bulking at all occurred when these ratios were less than 19:1 and 81:1 respectively, but there was some bulking at higher ratios. The bulking sludges contained *Sphaerotilus*.

NITRIFICATION

At most activated sludge plants, it is considered uneconomic to carry the purification of sewage beyond the clarification and carbonaceous oxidation stages. However, at some of the newer diffused-air plants (e.g. Mogden, West Hertfordshire (Colne Valley) and Hogsmill Valley) power is generated from the gas from the anaerobic digestion of the sewage sludge, which tends to make it an economic proposition to achieve a considerable degree of nitrification.

Lockett[27] has stressed the many benefits to be obtained by nitrifying sewage in the activated sludge process. Among these are
1. The production of a dense sludge, and hence, a large reduction in the amount of surplus activated sludge to be disposed of

Aerobic Biological Treatment

Table 14 (continued)

Activated sludge

Advantages	Disadvantages
Low	—
—	High*
Works well in winter in wet weather	Difficulties in dry summer months
—	Requires much
—	Rather sensitive to shocks of strong trade wastes
Little or no pumping needed since only a low head is required	—
Small	—
None	—
Suspended solids low	Usually not so highly nitrified as a filter effluent†
—	Large in quantity
—	Much foam, especially on air diffusion plants

* May be lower if all the sludge is digested and power is obtained from sludge digester gas
† Sewages containing inhibitory trade wastes are difficult to nitrify with activated sludge

2. Freedom from bulking
3. Plant is better able to deal with overloads of sewage and trade wastes and to work under adverse conditions
4. Sludge is more easily dewatered
5. The nitrified effluent, with its high nitrate content and low concentration of ammoniacal nitrogen*, is superior to an un-nitrified effluent in preserving good stream conditions; the nitrate is particularly useful in preventing bad stream conditions in industrial rivers containing organic pollution
6. The presence of nitrate in the effluent tends to inhibit foaming caused by synthetic detergents

At some activated sludge plants, part or the whole of the effluent is nitrified by treatment in percolating filters, e.g. at Oldham and Bury. On the other hand, some highly industrial sewages (e.g. at Manchester) appear to be very resistant to nitrification.

The following factors have an important influence on nitrification by activated sludge:

(i) High dissolved oxygen concentrations and long aeration periods favour nitrification

---

* Ammoniacal nitrogen is very toxic to fishes.

(ii) A high nitrogen content in the sewage (i.e. B.O.D.:N ratio less than 16) favours nitrification. Nitrification ceases if this ratio is 16 or more
(iii) Too low a concentration of activated sludge in mixed liquor adversely affects nitrification
(iv) The optimum pH for nitrification is 7·5–8·5. Nitrification slows down outside that range
(v) Nitrification is inhibited or prevented by small amounts of toxic substances from trade wastes, e.g. heavy metals, cyanides, many herbicides and insecticides, thiourea, and many other synthetic organic compounds.

COMPARISON OF PERCOLATING FILTERS AND THE ACTIVATED SLUDGE PROCESS

For the production of a Royal Commission effluent, it is generally necessary to choose between percolating filters and the activated sludge process. Important factors which may affect the choice are cost, local conditions and character and strength of the sewage. Where suitable land is scarce, e.g. near many large cities, the activated sludge process will be the obvious choice, since it needs a relatively small site area and is free from nuisance due to flies and odours. A percolating filter plant will occupy roughly seven times the area of an activated sludge plant to treat the same amount of sewage. Where plenty of land is available and a difficult industrial sewage is being handled, filters may be the better choice, as in the case of Bradford, Yorkshire, where a sewage containing much wool-scouring waste is dealt with by chemical treatment with sulphuric acid to remove wool greases, followed by biological filtration. When filters are used, a saving in capital expenditure as well as in area of land required can often be achieved by adopting alternating double filtration or recirculation instead of single filtration. This is being done at Birmingham, Liverpool, Bedford and many other towns in the British Isles. A combination of the two processes can also be used, e.g. activated sludge followed by percolating filters.

As a rough guide to assist in the choice of process, *Table 14* shows some of the advantages and disadvantages of percolating filters and the activated sludge process.

OXIDATION PONDS[28]

Oxidation ponds are large shallow basins, usually about 0·9–1·5 m (3–5 ft) deep, receiving a continuous flow of raw and settled sewage or even purified effluent. The minimum detention period may be some months for raw sewage and about 25–30 days for settled sewage. During this time, aerobic biological stabilisation takes place under the influence first of bacteria, then of algae. The organic matter is oxidised to carbon dioxide, ammonia (or nitrate) and water. The carbon dioxide is then utilised by algae to give oxygen in the presence of sunlight, and new algal cells are formed. The method has proved very successful in semi-tropical areas in the U.S.A. (e.g. California, Nevada and Arizona) and in Australia where land is cheap and the climate favourable (i.e. much sunshine to promote photosynthesis, low rainfall and relatively high air temperature), but use of these ponds has also been reported in more temperate climates (e.g. Sweden and Denmark). Oxidation ponds are used in the large new sewage treatment plant for Auckland (New Zealand). It is stated that ponds can be run for many years before removal of deposited sludge is necessary.

If properly operated, the ponds are resonably free from bad smells, due possibly to the deodorising effect of the chlorophyll in the algae. Treatment of sewage in an oxidation pond usually gives a good reduction in B.O.D., but the determination should be done on the filtered or centrifuged sample in order to remove the green algal cells. Remarkably good reductions of coliform bacteria are attained, usually over 99·5 per cent removal.

Algae from oxidation ponds are rich in protein and fat, and it has been proposed in the U.S.A. to use them as a cheap animal feeding stuff.

Although oxidation ponds are cheap to construct and operate, objectionable odours develop if they are overloaded, and they are also liable to cause the breeding of mosquitoes which may help to spread disease. The use of artificial aeration (e.g. by turbine aeration) may be necessary if existing installations are overloaded; they are then termed 'aerated lagoons' and resemble activated sludge plants since the floc particles contain ciliated protozoa but no algae[29].

Their use in this country is likely to be confined to the tertiary treatment of purified sewage effluents (see Chapter 7).

REFERENCES

1. *Royal Commission on Sewage Disposal*, 1901–15, 9 *Reports* with numerous Appendices; also a *Final Report*, 1915. London; H.M.S.O.
2. ARDERN, E. and LOCKETT, W. T., Experiments on the oxidation of sewage without the aid of filters, *J. Soc. Chem. Ind., Lond.* 33 (1914) 523; reprinted *J. Proc. Inst. Sew. Purif.* 3 (1954) 175
3. CHIPPERFIELD, P. N. J., The use of plastic media in the biological treatment of sewage and industrial wastes, *Surveyor, Lond.* 127 (1966) No. 3846, 30
4. WISHART, J. M. and WILKINSON, R., Purification of settled sewage in percolating filters in series with periodic changes in the order of the filters. Results of operation of experimental plant at Minworth, Birmingham, 1938–40, *J. Proc. Inst. Sew. Purif.* (1941) 15
5. DEKEMA, C. J. and MURRAY, K. A., Deep enclosed artificially ventilated filter beds versus ordinary open beds, *J. Proc. Inst. Sew. Purif.* (1942) 311
6. TEDESCHI, F. B. and LUCAS, R. W., Enclosed natural draught filter beds, *J. Proc. Inst. Sew. Purif.* 2 (1956) 161

7. LUMB, C. and EASTWOOD, P. K., The recirculation principle in filtration of settled sewage, *J. Proc. Inst. Sew. Purif.* 4 (1958) 380
8. Symposium on the evolution and development of the activated sludge process of sewage purification in Great Britain (9 papers) *J. Proc. Inst. Sew. Purif.* 3 (1954) 189–272
9. COOMBS, E. P., Air diffusers: their history and use in the activated sludge process, *J. Proc. Inst. Sew. Purif.* 4 (1955) 304
10. DOWNING, A. L. BAYLEY, R. W. and BOON, A. G., The performance of mechanical aerators, *J. Proc. Inst. Sew. Purif.* 3 (1960) 231
11. MCNICHOLAS, J. and TENCH, H. B., A review of recent activated sludge research at Manchester, *J. Proc. Inst. Sew. Purif.* 4 (1959) 425
12. PASVEER, A., Research on activated sludge. II. Experiments with brush aeration, *Sewage ind. Wastes* 25 (1953) 1397
13. TODD, J. P., Results from the modified Kessener plant at Rochdale, *J. Proc. Inst. Sew. Purif.* 5 (1963) 455
14. HIRST, J., Experimental work with a bio-aeration activated sludge plant, *J. Proc. Inst. Sew. Purif.* 6 (1964) 555
14a. BRIGGS, R., JONES, K. and OATEN, A. B., *Monitoring and Automatic control of dissolved oxygen level in activated sludge plants*, 1967. London; Effluent and Water Treatment Convention; also *Wat. Pollut. Abstr.* 41 (1968) p. 459
15. HASELTINE, T. R., Operating control tests for the activated sludge process, *Wat. Wks. Sewer.* 84 (1937) 121
16. MCCABE, J. and ECKENFELDER, W. W., JUN., *Biological treatment of sewage and industrial wastes.* Vol. 1. *Aerobic oxidation*, 1956, New York; Reinhold; see also GOULD, R. H., Sewage disposal problems in the world's largest city, *Sewage ind. Wastes* 25 (1953) 155
17. SETTER, L. R., CARPENTER, W. T. and WINSLOW, G. C., Practical applications of principles of modified sewage aeration, *Sewage Wks J.* 17 (1945) 669
18. MÖHLE, H., The high-rate activated sludge process at the sewage treatment Works at Wuppertal-Buchenhofen, *J. Proc. Inst. Sew. Purif.* 3 (1956) 297
19. *Ministry of Housing and Local Government, Extended aeration sewage treatment plants*, Circular No. 59/65, London; H.M.S.O. 1965
19a. STORCH, B., The design, construction and operation of extended aeration plants. *Wat. Pollut. Control* 68 (1969) 40
20. DOWNING, A. L., TRUESDALE, G. A. and BIRKBECK, A. E., some observations on the performance of extended-aeration plants. *Surveyor*, 124 (1964) No. 3771, 29
21. HOPWOOD, A. P. and DOWNING, A. L., Factors affecting the rate of production and properties of activated sludge in plants treating domestic sewage, *J. Proc. Inst. Sew. Purif.* 5 (1965) 435
22. SIMPSON, J. R., *Some aspects of biochemistry of aerobic organic waste treatment*, from pp. 1–30 of *Waste Treatment*, ISAAC, P. C. G. (Ed.), 1960. Oxford; Pergamon Press
23. ARDERN, E. and LOCKETT, W. T., Laboratory tests for ascertaining the condition of activated sludge, *J. Proc. Inst. Sew. Purif.* 1 (1936) 212
23a. CURDS, C. R., COCKBURN, A. and VANDYKE, J. M., An experimental Study of the role of ciliated protozoa in the Activated Sludge Process. *Wat. Pollut. Control.* 67 (1968) 312
24. KESSLER, L. H. and NICHOLS, M. S., Oxygen utilisation by activated sludge, *Sewage Wks J.* 7 (1935) 810
25. Standard methods for the examination of water and waste water, 12th Ed., 1965, New York; American Public Health Association
26. HATTINGH, W. H. J., Activated sludge studies, 3. Influence of nutrition in bulking. *Wat. Waste Treat. J.* 9 (1963) 476
27. LOCKETT, W. T., A contribution to the literature relating to the activated sludge process, *J. Proc. Inst. Sew. Purif.* 2 (1937) 88
28. *Review of sewage treatment processes and practice overseas and in New Zealand*, Pollution Advisory Council Publication No. 2, 1956. Wellington
29. MCKINNEY, R. E. and EDDE, H., Aerated lagoon for suburban sewage disposal. *J. Wat. Pollut. Control Fed.*, 33 (1961) 1277

Chapter 7

METHODS OF IMPROVING FINAL EFFLUENTS

INTRODUCTION

The normal standards proposed by the Royal Commission on Sewage Disposal for final sewage effluents discharging to inland watercourses, namely 5-day B.O.D. not greater than 20 mg/l and suspended solids not more than 30 mg/l, have not proved altogether satisfactory with the passage of time. They were intended to apply where the dilution available in the receiving stream was 8–150 volumes. But, in the years since the World War II, population has grown considerably, trade wastes discharged to the sewers have increased in volume and have become more and more complex in character and composition, there have been increases in the volume of sewage produced per head of population, and public water suppliers, industry and agricultural interests have abstracted larger amounts of river water. As a result, the minimum flows of many river waters have fallen and the proportion of effluent to river water has risen. In some instances we even have the unusual situation in which, far from there being 8 volumes of river to dilute 1 volume of effluent in dry weather, the effluent actually exceeds the river water in volume and so 'dilutes' the river—a state of affairs hardly foreseen by the Royal Commission. To be quite fair, however, the Royal Commission did, in Volume 1 of the 8th Report, envisage exceptional circumstances where, owing to the relatively small dilution afforded by the receiving stream, standards stricter than the normal ones might be imposed, though these were not defined or indicated but were to depend on the local circumstances.

Since the days of the Royal Commission, many factors other than dilution have been recognised to be of major significance in the re-aeration and self-purification of streams, particularly depth, temperature, degree of turbulence, and the character of the stream bed. Thus, a shallow, fast-flowing stream with a gravelly or rocky bed will undergo re-aeration and self-purification in a much shorter time than a stream which is deep, sluggish and has a bed of fine silt or mud. So a stream with a B.O.D. of 4 mg/l (the limiting value set by the Royal Commission) is not necessarily likely to cause nuisance; indeed, many good fishing rivers used for water supply have a five-day B.O.D. exceeding 4 mg/l. In a Memorandum issued with an explanatory Circular in 1966[1] by the Ministry of Housing and Local Government, it is stressed that although in many instances stricter standards than those of the Royal Commission may be imposed by River Authorities, such standards must be justified by local circumstances, and other factors in addition to dilution must be taken into consideration (e.g. depth and turbulence). On the other hand, where the Ministry has accepted that stricter standards are needed, the Local Authority must be prepared to meet the higher costs involved.

In recent years, several instances have occurred where the Ministry has agreed to the imposition of better standards than Royal Commission. These have mainly been where the receiving stream is used for public water supplies. For instance, at the Rye Meads Sewage Works (Middle Lee Regional Drainage Scheme)[2], part of the effluent is percolated into gravel strata to recharge the ground water. but the remainder, discharging to the River Lee, will have to pass the following standards:

Months	Not exceeding	
	B.O.D. mg/l	Suspended solids mg/l
May-October	5	5
November–April	10	10

These standards are unusually strict, since the River Lee is used for the water supply of London. It is to be noted that during the lower temperatures of the winter when biological purification of sewage is slower and it should be more difficult to produce good effluents, the standards are relaxed.

An obvious way of getting a high-quality effluent is to construct a plant of greater capacity than usual. This has been done at Crawley Sussex where the activated sludge plant has to give an effluent, not exceeding 15 mg/l in B.O.D. and 15 mg/l in suspended solids. There are various other methods and polishing devices for obtaining

Methods of Improving Final Effluents

high-quality effluents and these will be briefly reviewed in this chapter.

Many of the methods for improving final effluents often, referred to as the 'tertiary treatment' of sewage[1a], involve the elimination of suspended matter (fine humus or activated sludge) which is difficult to remove by ordinary settlement. Since the suspended matter in such effluents is largely organic, it is evident that removal of suspended solids will also lead to improvements in B.O.D., 4 h permanganate value and albuminoid (or organic) nitrogen [3].

MECHANICAL FLOCCULATION

Mechanical flocculation (cf. Chapter 5) has been used in the U.S.A. for the clarification of raw sewage as well as for the improvement of sewage effluents[4]. The process may be useful with humus tank effluents containing finely divided suspended matter which does not easily settle out in the tanks. Hurley and Lester[5] carried out laboratory tests on different humus tank effluents, using flocculation times of 15–30 min followed by about 30 min quiescent settlement. The paddle speeds were quite low, 15 rev/min, equivalent to a peripheral speed of 0.07 m/s (0.23 ft/s). The reduction of suspended solids was often of the order 50–60 per cent, although in some instances there was practically no improvement. The flocculated settled effluents contained about 6–17 mg/l of suspended solids.

The process, which is continuous and would require little attention, does not seem to have been much used in Great Britain for the improvement of effluents.

MICRO-STRAINERS

A micro-strainer is a drum of woven stainless steel fabric (with extremely minute orifices) rotating on a horizontal axis and partly submerged in the liquid to be filtered. Micro-strainers were first used in waterworks practice but experiments by the Water Pollution Research Laboratory led to their introduction at the Luton Sewage Works in 1950, which was a new development in sewage purification. The strainers used at Luton to remove suspended matter have 100 000 orifices 645/mm^2 (100 000 orifices/in^2) of 0.045 mm diam. As a result of a common law action, Luton had the formidable task of producing from an industrial sewage an effluent approximating in quality to a river water. Treatment consists of sedimentation, activated sludge treatment and biological filtration. Part of the humus tank effluent then receives sand filtration (see below), part goes through micro-strainers, three of which treat about $11\cdot27 \times 10^6$ l/day ($2\cdot5 \times 10^6$ gal/day). Average results for the micro-strainers for the year 1956 were[6]

	Humus tank effluent	Micro-strainer effluent	Reduction by micro-strainers per cent
B.O.D. (mg/l)	9·2	6·6	28
Suspended solids (mg/l)	13·4	6·3	53

It is necessary to 'backwash' the micro-strainers with high-pressure jets of the filtered effluent; these wash waters are returned to the sewage inlet for treatment. The amount of effluent used for backwashing is about 3 per cent of the throughput.

At the Hazelwood Lane Sewage Plant of the Bracknell Development Corporation, where the Thames Conservancy Board requires an effluent with a suspended solids figure of not more than 10 mg/l, micro-strainers gave the following results for the year ending 31st March, 1958[7]:

	Humus tank effluent	Micro-strainer effluent	Reduction by micro-strainers per cent
B.O.D. (mg/l)	14·7	10·6	28
Suspended solids (mg/l)	15·2	6·5	57

A sample of the micro-strainer wash waters from this works contained 717 mg/l of suspended solids.

A biological film forms on the mesh of micro-strainers, which has to be removed from time to time. This can be done by washing the mesh with a solution of chlorine. The interval between washings varies from 7–28 days, depending on the character of the sewage effluent. The formation of this biological film can also be prevented by the use of ultra-violet lamps.

Micro-strainers have the advantage over sand filters of requiring only very small areas of land; they are also cheaper to run. The maximum head loss is usually about 300 mm (12 in), and they require very little attention.

If micro-strainers are to be operated efficiently there is a limit to the amount of suspended solids which can be permitted in the effluent to be passed through the micro-strainer. The amount of colloidal matter present has also a considerable effect on the operation of these strainers, and the less colloidal matter present the higher will be the permissable level of suspended solids in the incoming liquid. Boucher[8] points out that the limit of suspended sewage humus solids may be between 40 and 60 mg/l, but adds that one cannot be dogmatic about this as much depends on prevailing local conditions.

SAND FILTERS

The use of rapid gravity sand filters for the removal of fine suspended matter from sewage effluents has been tried at a few places in the British Isles. At the Luton Sewage Works, where they are operated in parallel with micro-strainers[6], there are nine rapid gravity sand filters operating at 9 800 l/m² h (200 gal/ft² h). Average results for the year 1956 were

	Humus tank effluent	Sand filter effluent	Reduction by sand filters per cent
B.O.D. (mg/l)	9·2	4·0	57
Suspended solids (mg/l)	13·4	3·3	71

It was found necessary to backwash the sand filters once or twice a day, the washings being returned to the sewage inlet.

Comparing these figures with those for micro-strainers, it can be seen that the latter gave rather lower reductions in B.O.D. and suspended matter. It would appear that some biological action takes place in the sand filters, so the effluent is better as regards B.O.D. and suspended matter, but contains less dissolved oxygen; this is no disadvantage, as re-aeration of the River Lee occurs within a short distance.

Sand filters are used for 'polishing' the activated sludge plant effluent at the Rye Meads Sewage Works.

Sand filters give good removal of bacteria, provided they are operated at slow rates—certainly much slower than those used at Luton. In South Africa, experiments have shown that poliomyelitis virus, amoebic cysts and worm eggs (which commonly occur in tropical sewages) are removed from sewage effluents by slow sand filtration, and this treatment is now given to all sewage effluents in the Union before they are used for irrigation.

Slow sand filters consist of basins containing 0·3–0·76 m (1–2½ ft) of sand on coarser material resting on an underdrainage system. They are simpler in construction than rapid gravity sand filters and, therefore, can be used at small sewage works requiring a high-quality effluent. No backwashing was carried out on the early types of these filters, but the surface layer had to be removed and replaced by fresh sand when the loss of head became too great. Later types have been adapted to allow backwashing to be carried out[9].

UPWARD FLOW FILTERS

The conventional filters used in the past have been downward flow or gravity filters. A new development is the use of upward flow filters of which two are already finding increasing application.

IMMEDIUM SAND FILTER

This is an upward flow sand filter about 1·5 m (5 ft) deep available from Messrs W. Boby and Co., Rickmansworth, Herts ('Boby-Imatic Immedium Filter'). The sand rests upon a grid and the particles after agitation naturally separate so that the larger grains are at the bottom and the smaller grains are at the surface. The effluent being filtered, pumped under pressure from the bottom, runs first through the coarser sand particles, which remove the larger suspended solids, and then through the smaller sand particles, where the finer solids are retained. Thus, the effluent receives a progressive degree of polishing as it passes from the bottom upwards. With this upward flow filter, therefore, the whole of the filter bed is utilised, while with the older downward flow type of filter, only the top layer of sand was fully utilised. In practice, this means not only improved filtration results and higher flow-rates but also much longer runs between the filter washes.

In tests with a feed liquor containing 33 mg/l of suspended solids treated at the rate of 9 800 l/m^2 h (200 gal/ft^2 h)[10], the Immedium filter gave a final effluent with only 6 mg/l of suspended matter whereas the conventional downward flow sand filter effluent contained 13 mg/l.

Methods of Improving Final Effluents

BANKS UPWARD FLOW GRAVEL CLARIFIER[11,12]

Another very promising filter for removing suspended solids from biologically treated effluents has been developed by Banks[11]. It can be incorporated in an existing humus tank as indicated in *Figure 22*.

It consists of a bed of pea-gravel (i.e. small shingle) about 6–10 mm ($\frac{1}{4}$–$\frac{3}{8}$ in) in diameter and about 150 mm (6 in) in depth. The gravel has, of course, to receive some form of support, e.g. a perforated floor or tray which can be of steel or galvanised iron mesh or of perforated pre-cast concrete. The upward flow is at the rate of about 980 l/m²h (20 gal/ft²h). Solids accumulating in the bed are removed periodically

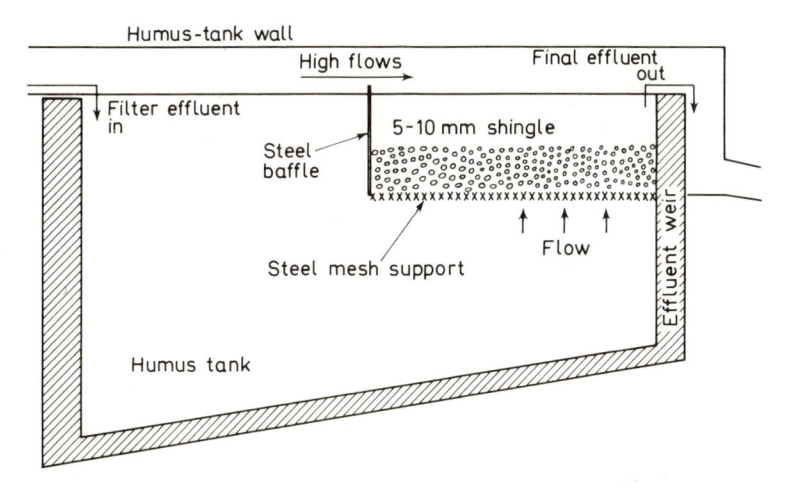

Figure 22. Banks upward-flow pebble clarifier installed in a horizontal-flow humus tank. (Reproduced from J. Proc. Inst. Sew. Purif. 1 (1966) 45 by courtesy of the Editor)

by backwashing, i.e. lowering the water level by allowing effluent to flow out below the gravel bed; any further cleaning may be done with a jet of water or effluent.

The Banks clarifier affords a simple and cheap method of removing suspended solids and giving a final effluent with as little as 6–8 mg/l of suspended matter.

CLARIFICATION LAKES OR LAGOONS

It is well known that sewage effluents when stored in large shallow artificial lakes undergo further purification by micro-organisms in the presence of dissolved oxygen (cf. oxidation ponds, p. 106). It has been extensively used abroad in warm sunny climates e.g. in California, Israel, and South Africa. It is much used in South Africa where there is a great shortage of clean water. Hence, sand filtered sewage effluents are given further treatment in so-called 'maturation ponds' so as to achieve a high quality effluent of extremely low bacterial count.

The method is beginning to find application in England. An artificial lake for the clarification of humus tank effluent is now in use at Witham U.D.C., Essex[13]; the lake is about 0·9 m (3 ft) deep and has a total capacity of about $4 \cdot 1 \times 10^6$ l (900 000 gal), thus giving about 3 days retention.

At Dunstable[14], where standards stricter than Royal Commission are required, the humus tank effluent receives tertiary treatment in four lagoons arranged in series which give an excellent final effluent (B.O.D., 3·8 mg/l; suspended solids, less than 1 mg/l).

The method requires the use of rather large areas of land and is, therefore, only suitable in this country for small sewage works.

LAND TREATMENT

Provided that suitable land is available, land treatment is a satisfactory method of improving sewage effluents. Indeed, the Severn River Board has proposed irrigation over grassland as a simple and economic method of obtaining high-quality sewage effluents.

For some years, the Upper Tame Main Drainage Authority and its predecessor have successfully used grass plots for improving the humus tank effluent from the Barston Works[15]. Results obtained during 1956 were:

	Humus tank effluent	*Grass plot effluent*
B.O.D. (mg/l)	24·8	10·5
Suspended solids (mg/l)	48·0	12·0
Ammoniacal nitrogen (mg/l)	8·8	6·6
Oxidised nitrogen (mg/l)	22·6	14·9

It will be noticed that good reductions in B.O.D. and suspended solids (bringing the effluent within Royal Commission Standards) contrast with a marked fall in oxidised nitrogen content, probably due to utilisation by soil bacteria and by the growing grass.

Dosages up to 8.4×10^6 l/hectare day (0.75×10^6 gal/acre day) on grass plots can bring about a considerable improvement in the quality of sewage effluents, but smaller dosages are generally used. The land must be of a suitable nature (peat and clay lands are not to be recommended) and should preferably have a suitable fall, but not steeper than 1 in 60.

NITRIFYING FILTERS

Nitrifying filters can be used with advantage to improve the effluents from activated sludge plants, which generally contain very little nitrate but can have a fairly high content of ammoniacal nitrogen. If such activated sludge plant effluents are discharged into streams affording very little dilution there can be a reduction in the dissolved oxygen of the stream water due to it being used to oxidise the ammoniacal nitrogen. The presence of the ammoniacal nitrogen in the effluent can also have an adverse effect on any fish life in the receiving stream. If the effluent is to be discharged into a stream which receives a heavy pollution load from other sources, the presence of nitrate in the effluent would play its part in preventing septic conditions being set up in the stream water. Hence, the use of nitrifying filters in these circumstances can be of great benefit. In addition they can effect a considerable reduction in the B.O.D. and produce an effluent much better than Royal Commission standard.

For many years a nitrifying filter which was only 0.9 m (3 ft) deep was operated at the Bury Corporation Sewage Works. The dosage of activated sludge plant effluent on the filter was 1 800 l/m^3 day (300 gal/yd^3 day). This effected a reduction in the B.O.D. of over 30 per cent and the nitrate content of the effluent always exceeded 10 mg/l.

REMOVAL OF NUTRIENTS

Sewage effluents inevitably contain such nutrients as phosphates and nitrates and when the effluents are discharged into free-running streams, giving adequate dilution, and not used for drinking water purposes, the presence of these constituents does not normally present any problems. However, when the effluents are discharged into streams which are used to supply drinking water their presence tends to be objectionable. If the effluents are discharged into lakes, ponds or sluggish streams these nutrients tend to stimulate growths of phytoplankton ('algal blooms'), and can also stimulate weed growth. This can lead to 'eutrophication' of the lake, an ageing process due to excessive discharge of nutrients. Eventually, this may go far enough to convert the lake to marshland or even dry land as has happened in the U.S.A. Under such circumstances it may become necessary to take steps to remove nutrients from the effluents. To this end Bush[16] and his associates used an algal lagoon, in West Los Angeles, to treat an activated sludge plant effluent. The nutrients were removed by algae growing in the pond. The predominant alga was *Scenedesmus* which attaches itself to the bottom and sides of the lagoon and can thus be removed by scraping after the lagoon has been dewatered. It has been found, however, that when this is carried out in places which have cold winters, the phosphate removal in winter is poor and the process is not very reliable[17,18]. Owen[19] found by laboratory trials that if lime is added to a sewage effluent until the pH is 11·0, almost complete removal of phosphate as insoluble $Ca_3(PO_4)_2$ takes place. When lime was added to the influent of a humus tank, there was a reduction of 77 per cent in the phosphorus content. Unfortunately this method produces a strongly alkaline effluent and produces large volumes of sludge. Lea, Rohlich and Katz[20] showed that 77–89 per cent of phosphate in an effluent could be removed by coagulation with alum or sodium aluminate (200 mg/l). Bringmann[21,22] showed that nitrogen can be removed from a nitrified effluent by mixing it with crude sewage at an optimum redox potential; the resulting effluent being finally aerated to oxidise any remaining organic compounds.

REFERENCES

1. *Ministry of Housing and Local Government, Technical problems of River Authorities and Sewage Disposal Authorities in laying down and complying with limits of quality for effluents more restrictive than those of the Royal Commission*, London; H.M.S.O. 1966. See also Circular No. 37/66
1a. TRUESDALE, G. A. and BIRKBECK, A. E., Tertiary treatment processes for sewage works effluents, *Wat. Pollut. Control* 66 (1967) 371
2. ANON., The Middle Lee Regional Drainage Scheme, *Surveyor, Lond.* 116 (1957) 1053
3. STONES, T., Removal of suspended solids from filter effluents, *J. Proc. Inst. Sew. Purif.* 1 (1956) 107
4. FISCHER, A. J. and HILLMAN, A., Improved sewage clarification by pre-flocculation without chemicals, *Sewage Wks J.* 12 (1940) 280.
5. HURLEY, J. and LESTER, W. F., Mechanical flocculation in sewage purification, *J. Proc. Inst. Sew. Purif.* 2 (1949) 193
6. EVANS, S. C., Ten years operation and development at Luton Sewage Treatment Works, *Wat. Sewage Wks* 104 (1957) 214
7. *Bracknell Development Corporation, 2nd Annual Report for period ending 31st March, 1958*. Bracknell, Berkshire
8. BOUCHER, P. L., Micro-Straining, *Inst. Publ. Hlth Eng J.* 60. (1961) No. 6, 118
9. HUTCHINGS, A., Report on tests carried out on the operation of sand filters of small sewage works at Horsted Keynes and Handcross, *J. Proc. Inst. Sew. Purif.* 3 (1957) 252

10. ROBERTS, F. W., Developments in tertiary treatment of effluents, *Munic. Engng, Lond.* 142 (1965) 2707
11. BANKS, D. H., The development of a clarifier for use in treating sewage effluents, *Wat. Waste Treat. J.* 10 (1964) 24; 10 (1965) 241. Also *Surveyor, Lond.* 123 (1964) No. 3745, 21; 125 (1965) No. 3789, 45
12. TRUESDALE, G. A., BIRKBECK, A. E., and DOWNING, A. L., The treatment of sewage from small communities, *J. Proc. Inst. Sew. Purif.* 1 (1966) 34
13. ANON., Lake settlement of effluent, *Effl. and Wat. Treatment J.* 3 (1963) No. 1, 16
14. ANDREWS, L., Operational experiences at Dunstable sewage works: system of double filtration and recirculation in plant constructed 1955–58, *Surveyor, Lond.* 123 (1964) No. 3739, 32,56; also *J. Proc. Inst. Sew. Purif.* 5 (1964) 434
15. DAVISS, M. R. V., Treatment of humus tank effluent on grass plots, *Surveyor, Lond.* 116 (1957) 613
16. BUSH, A. F., ISHERWOOD, J. D. and RODGI, S., Dissolved solids removal from waste water by algae, *Proc. Am. Soc. civ. Engrs* 87 (1961) SA3, 39–57, Pap. No. 2824, also *Wat. Pollut. Abstr.* 35 (1962) 168–9 and *J. Wat. Pollut. Control Fed.* 34 (1962) 450
17. MACHENTHUN, R. M. and MCNABB, C. D., Stabilisation pond studies in Wisconsin, *J. Wat. Pollut. Control Fed.* 33 (1961) 1234–51
18. ROHLICH, G. A., Methods for the removal of phosphorus and nitrogen from sewage plant effluents, *Int. J. Air Wat. Pollut.* 7 (1963) 427–34
19. OWEN, R., Removal of phosphorus from sewage plant effluents with lime, *Sewage ind. Wastes* 25 (1953) 548–56
20. LEA, W. L., ROHLICH, G. A. and KATZ, W. J., Removal of phosphates from treated sewage, *Sewage ind. Wastes* 26 (1954) 261–75, see also *Int. J. Air Wat. Pollut.* 8 (1964) 487–500
21. BRINGMANN, G., Optimum removal of nitrogen gas by addition of nitrifying activated sludge and redox control, *Gesundheitsing enieur* 81 (1960) 140–2; also *Wat. Pollut. Abstr.* 34 (1961) 310
22. BRINGMANN, G., Complete biological elimination of nitrogen from clarified sewage in conjunction with a high-efficiency nitrification process, *Gesundheitsing enieur* 82 (1961) 233–5; also *Wat. Pollut. Abstr.*, 35 (1962) 345

Chapter 8

SLUDGE TREATMENT AND DISPOSAL

INTRODUCTION

The purification of sewage results in the production of large volumes of semi-liquid sludges, and these have to be disposed of without causing nuisance, pollution of watercourses, danger to public health or damage to public amenities. This presents a very difficult problem, especially when it is realised that a national average of 25 tons of dry solids[1] are produced from the sewage from 1 000 persons per annum. The actual amount produced varies greatly from works to works, depending on the type of treatment processes used, the nature of the sewage and local conditions.

There are three main types of sludge which can be produced on a sewage works:

(a) Primary sludge, which consists of the solid matter settled out in the primary sedimentation tanks. According to Escritt[2] it can be assumed that the sludge produced when settlement of domestic sewage takes place in hopper-bottomed tanks is 2·3 l/day ($\frac{1}{2}$ gal/day) per head of population (moisture content about 97·5 per cent), with mechanically cleaned tanks 1·1 l/day ($\frac{1}{4}$ gal/day) (moisture content about 95 per cent). This corresponds with about 0·057 kg (0·125 lb) of dry solids per head of population per day. But the actual amount produced will vary from works to works and is dependent on the suspended solids content of the raw sewage, the settlement period and the efficiency of the settling tanks. The nature of primary sludge varies, depending on the nature of the sewage and especially on its trade waste content. The presence of toxic metals can have an adverse effect on the use of the sludge as a fertiliser. Wastes from food processing factories and breweries can cause the sludge to ferment, and tripe-dressing and wool-scouring wastes will result in the production of a sludge with a high grease content. Trade wastes having a very high suspended solids content, such as paper mill wastes, will, if not

Sludge Treatment and Disposal

properly pre-treated before discharge to the sewers, greatly increase the volume of sludge produced. Where chemical coagulation is used to assist sedimentation, likewise a greater volume of sludge results containing more mineral matter than sludge from plain sedimentation, and having a lower moisture content.

(b) Humus sludge, which consists of the solid matter settled out in humus tanks from the effluent of percolating filters. This sludge, with a moisture content of 94–97 per cent, is rather difficult to dewater. It is greyish-brown in colour, and when fresh the odour is not unpleasant. According to Imhoff et al.[3], about 0·013 kg (0·029 lb) of dry humus solids are produced by conventional low-rate percolating filters per head of population per day. During the warmer weather of spring and early summer, the volume of humus sludge produced increases considerably due to 'sloughing off' or 'unloading' from the filters when the biological film is attacked and broken up by the macro-organisms.

(c) Activated sludge is the surplus sludge from an activated sludge plant. It is brownish, flocculent and has a pleasant earthy smell when fresh, but if allowed to become septic acquires a very offensive odour. It has a moisture content of 98–99·5 per cent and is difficult to dewater. It contains 5–7 per cent N and 2–4 per cent P_2O_5 (dry basis, respectively). Very large volumes are produced by activated sludge plants, thus creating a difficult disposal problem.

It is important to realise that sludges from different sewage works vary considerably in their physical properties. This may be due to differences in the food habits in different areas and to differences in the trade wastes present.

On many sewage works the surplus activated sludge and/or humus sludge are returned to the inlet of the primary sedimentation tanks and allowed to settle out with the primary sludge, producing a mixed sludge.

The sewage works manager is thus faced with the problem of satisfactorily disposing of large volumes of sludge which can have a moisture content varying from about 90 to 99·5 per cent. It is obviously desirable to reduce the moisture content to facilitate handling and disposal, and this will be evident from the following formula which gives approximately the change in volume of sludge by removal of water

$$\frac{V_1}{V_2} = \frac{S_2}{S_1} = \frac{100 - M_2}{100 - M_1}$$

where V_1 and V_2 = sludge volumes

S_1 and S_2 = dry solids content (percentages by weight)

M_1 and M_2 = moisture contents (percentages by weight)

It will be seen that the volume of a sludge varies inversely with the percentage of dry solids; thus a sludge of 94 per cent moisture occupies only one-third of the volume of a sludge of 98 per cent moisture content if both contain the same weight of dry solids. It is thus advantageous to reduce the moisture content as much as possible. In mechanically cleaned sedimentation tanks which have sludge hoppers of large capacity some consolidation takes place in the hoppers, otherwise consolidation tanks can be provided. These consist of tanks often fitted with slowly rotating picket fence thickeners; when the sludge has been allowed to settle the supernatant water is run off via the decanting valves.

As the moisture content of sludge is reduced to about 85 per cent a definite thickening is noticeable. At a moisture content of about 70–80 per cent, the sludge no longer flows and is known as 'sludge cake' in which condition it is easily handled with a spade. At low moisture contents (e.g. 10–15 per cent) it appears to be as dry as dust.

DISPOSAL OF LIQUID SLUDGE

SEA DISPOSAL

Wet sludge is collected in storage tanks adjacent to the landing stage from where it is run into tanks in specially designed ships and taken out to sea. When the vessel reaches an area of deep sea, valves are opened and the sludge is discharged into the sea while the ship is in motion. This method of disposal is practised at Manchester, Salford, London, Southampton and Glasgow. In the past this was considered an economic method of disposing of sludge, but owing to the present-day high costs of ships, labour, fuel and harbour dues, and the fact that ships are often tied up for appreciable periods of time, it is now regarded as costly. However, local conditions can be such that it is the only practicable and economic method of disposal. The method has, in the past, been only applicable for those communities which have convenient access to the sea, but improved methods of pipe-line construction have allowed inland local authorities to give serious consideration to this method of disposal.

DISPOSAL ON LAND

Where large areas of suitable land are available these can be used for disposal of liquid sludge, and various methods are available.
Trenching—Trenches are dug in the land and partially filled with

Sludge Treatment and Disposal

liquid sludge, then refilled with soil. This method is not usually practical for the larger works.

Land Plots—The land is ploughed and the sludge run into the furrows and allowed to dry, after which the land is re-ploughed and can be used for the growing of crops. This method makes use of the fertilising constituents of the sludge, and any risks of transmission of disease by pathogenic organisms are virtually ruled out if it is well ploughed in before cropping. Dressing of the land with lime will be necessary to prevent the development of acidity in the soil. According to Jenkins[4] the total area of land required is about 9 hectares/10^6 l (100 acres/10^6 gal) of daily D.W.F. of sewage, or about 22·6 m^2 (27 yd^2) per person.

Lagoons—These are virtually tips for liquid sludge, made by constructing high bankings round an area of land; natural depressions can sometimes be utilised, thus reducing the height of bankings required. Sludge is run into the lagoon. Provision is usually made for decanting off supernatant water which must be given suitable treatment before discharge into a watercourse; wherever possible it should be returned to the inlet of the sewage works. Digestion of the sludge takes place in the lagoon, with some reduction in volume. The sludge is never removed from this type of lagoon, and where suitable land is available this is a cheap method of disposal.

Lagoons can, however, be used as cold digestion tanks, and after a long period of digestion the sludge is run on to drying beds, or land to be dewatered.

Shallow lagoons can be used as drying beds, especially if they are underdrained.

In all these methods of disposing of sludge on land, if the land is underdrained this drainage may be polluting in character and should be given treatment before being discharged to a watercourse, preferably by being returned to the sewage works inlet.

COMPOSTING

Wet sludge can be composted with household refuse and also with straw and grass. In the former case, metal objects, glass, broken crockery, cinders and fine dust, large pieces of paper and cardboard are first removed. The composting can be carried out in shallow pits, bays or composting cells. The heap is built up in layers of household refuse, the sludge being run on to the layer before the next layer is laid on top. The ratio of sludge to refuse is usually in the region of 2 or 3 to 1 by weight, depending on the absorbent properties of the refuse, but higher ratios have been used. There must be adequate ventilation of the heap, and a drainage system must be provided underneath. Some of the liquid draining from the heap can be used to keep the material moist but any surplus must be returned to the inlet of the sewage works, as it is a most polluting liquid. It is usual to turn the material periodically; in some systems the mixing is carried out before placing it in the composting cell. In an efficient compost heap temperatures of about 70°C will be reached, which will result in the destruction of harmful organisms and weed seeds. The time for composting will vary from 3 to 6 weeks, depending on the efficiency of the method used. After this period the compost is placed in heaps, preferably under cover, and allowed to mature for at least three months. Organic material such as bracken or straw may have to be added to household refuse before composting to maintain a final carbon/nitrogen ratio of from 15–20 to 1 in the compost[1]. Sludge and refuse can thus be converted into a valuable fertiliser, but composting in this manner is generally considered to be uneconomic owing to the long period necessary to effect it efficiently and the heavy labour costs involved.

The development of mechanical equipment for composting has reduced the processing time and made composting a more economic proposition. In the Dano System which is used at Edinburgh[5], Renton, Leatherhead and Leicester, sludge and household refuse are rotated in a metal cylinder, and air is blown through the material from jets. The outlet end of the cylinder is slightly lower than the inlet end so that, in addition to being rotated, the material is slowly moving along the cylinder. Accelerated fermentation takes place and a temperature of about 65°C is attained in the cylinder. The process is completed in 5 to 6 days. It should be remembered that some industrial sewage sludges containing toxic metals may not be suitable for composting.

Municipal composts are a good source of those trace elements (copper, cobalt, zinc, manganese, iron, etc.) that are so necessary for satisfactory plant growth.

TREATMENT OF SLUDGE TO FACILITATE DEWATERING

CHEMICAL CONDITIONING

Chemical conditioners are sometimes added to sewage sludge as a form of preliminary treatment to facilitate drying on beds or filter pressing, and it is essential to add them before dewatering sludges on vacuum filters.

Apart from sulphuric acid, copperas and lime, which have special

Sludge Treatment and Disposal

uses, the commonly used chemicals are the compounds of the trivalent metals iron and aluminium. In order of effectiveness as sludge conditioners, starting with the most efficient, they are

Aluminium chlorohydrate*, $Al_2(OH)_4Cl_2$

Aluminium chloride†, $AlCl_3$

Ferric chloride†, $FeCl_3$

Chlorinated copperas‡, $Fe\begin{smallmatrix}\diagup Cl \\ \diagdown SO_4\end{smallmatrix}$

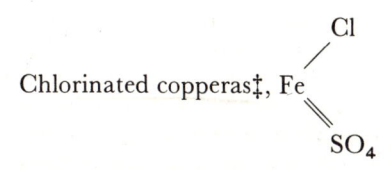

Ferric sulphate, $Fe_2(SO_4)_3$

Aluminium sulphate, $Al_2(SO_4)_3.18H_2O$

Sulphuric acid is used in the woollen towns of Bradford and Halifax to acidify primary tank sludges (to pH about 3·5) so as to liberate valuable wool greases from these highly greasy sludges before filter pressing.

Lime is used mainly to condition primary sludges ahead of filter pressing, the amount added being usually about 5–10 per cent, calculated on a dry sludge solids basis. Lime in conjunction with copperas was found to be the most economical combination of chemicals for conditioning digested sludge before dewatering on filter presses at the Wandle Valley Sewage Works[6].

In the U.S.A. lime has also been used in conjunction with ferric chloride to condition primary sludge before vacuum filtration, but when elutriation is practised the use of lime is unnecessary.

Iron salts—Ferric chloride is much used in the U.S.A. for coagulating sludges prior to vacuum filtration but it is no longer available in the British Isles in large quantities. At the works of the West Hertfordshire Main Drainage Authority (formerly Colne Valley), chlorinated copperas, used as a coagulant for the sludge before vacuum filtration, is prepared by passing chlorine into a solution of commercial copperas whereupon the divalent iron is oxidised to the trivalent state

$$6FeSO_4.7H_2O + 3Cl_2 = 2FeCl_3 + 2Fe_2(SO_4)_3 + 42H_2O$$

Aluminium salts—Aluminium chlorohydrate[7] has been shown to be superior to ferric salts as a sludge conditioner and is now used

---

* A chemical marketed by Laporte Industries Ltd., Widnes
† Not available commercially in large quantities in Great Britain
‡ Really a mixture of $FeCl_3$ and $Fe_2(SO_4)_3$.

at most plants using vacuum filtration for sewage sludge dewatering. It is also a useful compound for conditioning all types of sewage sludge which are to be dewatered on drying beds[8]. At Hamilton, where $AlCl_3$ is available as a waste product from a chemical plant, Stillingfleet[9] found it to be most effective as a conditioner for digested sludge prior to filter pressing.

Organic polymers—In the U.S.A., synthetic organic water-soluble polymers (e.g. Separan 2610, a high molecular weight polyacrylamide) have been introduced in recent years as flocculating aids, in conjunction with ordinary chemical coagulants, for conditioning sludge prior to vacuum filtration. It is claimed that the use of very small amounts of these polymers can reduce very considerably the ferric chloride dosage, thus resulting in overall savings in chemical material costs. Similar organic polymers are now used in Britain.

ELUTRIATION

Elutriation is the process of washing sewage sludges (especially primary and digested sludges) prior to coagulation with chemicals (such as ferric and aluminium salts) and subsequent vacuum filtration. This method serves to remove soluble ammoniacal compounds which consume and waste the coagulant, substances which reduce ferric salts to much less efficient ferrous salts, and finely divided suspended matter which interferes with filtration.

Elutriation is carried out in one or two stages by mixing the sludge with several volumes of water, purified plant final effluent or even settled sewage. This can be done either mechanically in tanks equipped with stirring and scraping devices or else with diffused air. The mixture is then allowed to settle for several hours and the supernatant liquor run off and returned to the sewage inlet for treatment.

Economy in wash waters can be secured by using 'counter-current' washing in which the weaker elutriates, before being returned to the sewage inlet, are used to elutriate a fresh batch of sludge. This procedure is suitable for the larger plants where it can be operated continuously in two tanks in series.

The chief advantage of elutriation is that it reduces very greatly (often by as much as 80 per cent) the coagulant demand and so results in considerable savings in chemicals. It also renders the use of lime unnecessary when filtering primary sludges in vacuum filters.

In sludge digestion plants, elutriation is sometimes used in the U.S.A. to 'clean up' the primary tank digested sludge, thus allowing considerable reduction in the capacity of the secondary digestion tanks.

FREEZING

It has been shown that alternate slow freezing and thawing of primary, digested and activated sludges alters the character of the sludges to such an extent that water separates out quite easily. The addition of conditioning chemicals (e.g. Al salts or chlorinated copperas) before freezing increases the subsequent filtration rate. Filtration of the sludge then gives cakes with as high a dry solids content as 23–30 per cent. The process is patented and, though not likely to be cheap, is being tried out on a working scale.

The principle of slow freezing is to form large ice crystals so that, on thawing, the water will separate easily from the solids. If rapid freezing were carried out small crystals would be formed, and on thawing the water would be taken up again in to the solid matter.

PORTEOUS PROCESS

This process, which owes its successful development to W. K. Porteous but is now in the hands of the firm Hawker Siddeley, Activated Sludge Ltd., is a method of conditioning the sludge by heat under pressure. The sludge is heated in steel cylinders with live steam at about 180°C and 103×10^4 N/m^2 (150 lb/in^2) pressure for 45 min or longer, which breaks down the gel structure of the sludge so that, after cooling, water separates out easily by settlement. The conditioned sludge can then be dewatered on filter presses. An important part of the plant is the heat exchanger, or 'coactor', which is essential to achieve a high heat recovery. Foul gases are produced by this process, which could be a nuisance to local residents. Steps may have to be taken to deal with these gases, as far as possible. Gases from the heater have been discharged under the boiler fires, where they are burned.

The process is particularly useful for secondary sludges which are difficult to dewater. It was used by Lumb[10] at the Halifax Sewage Works for many years for dealing with mixed humus and activated sludges. He obtained a filter press cake of about 50 per cent moisture content within 2 days at this plant from sludges containing about 95 per cent moisture. The sludge cake was quite sterile and was sold as manure. The liquor produced was extremely foul, with a B.O.D. of about 4 600 mg/l, and was dealt with by return to the sedimentation tanks balanced over the day.

The process is in use at Halifax, Huddersfield and Barrhead (Scotland), but plants at Horsham and Luton have discontinued operation because of high costs. At Halifax steam was economically produced in boilers heated with cinders supplied by the Cleansing Department. The process can readily be adapted to complete automatic operation[11].

Several firms now produce sludge conditioning plants utilising the heat treatment principle.

In all heat treatment plants the production of foul liquors presents a treatment problem due to the additional load they place on the sewage works, and in some cases it may be necessary to provide special pre-treatment plant to deal with these liquors. At the least they have to be balanced out over the day. The production of foul smelling gases also presents problems, especially if the sewage works is situated in a built-up area.

SLUDGE DIGESTION

Although sewage sludge (especially sedimentation tank sludge) undergoes fermentation or 'digestion' by anaerobic bacteria (Chapter 2) when stored in tanks for a long time, the process is slow at ordinary temperatures (about 3–6 months), and is used under these conditions only at the smaller works. A layer of scum which forms on the surface effectively prevents access of oxygen from the air, which would stop the process. The digestion process can be accelerated by raising the temperature of the sludge. Digestion carried out at about 30°C is called 'mesophilic digestion', and this is the type usually practised in Britain. Digestion carried out at a temperature around 50°C is known as 'thermophilic digestion'; the higher temperature speeds up digestion but is not easy to control, and foul odours and a very foul liquor are produced.

At the medium-sized and larger works it is becoming customary to carry out the digestion in two stages:

1. The primary stage is effected in heated closed tanks, provided with fixed or floating roofs, for 7–30 days at 27–35°C. Most of the digestion and gas evolution takes place here. Heating can be carried out by circulating the contents through an external heat exchanger, by means of internal coils containing hot water or even by injecting steam into the contents. Part of the gas evolved is utilised as a source of heat. Where the detention period is low (e.g. less than 20 days) the process is often called high-rate anaerobic digestion.
2. The secondary stage follows for 10–60 days at ordinary temperatures in open tanks, when some further digestion occurs and consolidation of the sludge takes place, with separation of much liquor.

Sludge Treatment and Disposal

Digestion results in a drop in the organic contents of the sludge, the end products being gas containing chiefly about 70 per cent methane (CH_4) and 30 per cent carbon dioxide (CO_2), and soluble ammoniacal compounds produced from the organic nitrogenous matter.

In a town of, say, 100 000 population with some trade waste, if we assume a figure of 2·3 l/head (½ gal/head) per day of sludge containing 97 per cent moisture, the daily sludge production will be 230 000 l (50 000 gal). If the sludge is densified before digestion to 95 per cent moisture, the volume passing to the digesters will be 137 000 l/day (30 000 gal/day). Allowing 30 days digestion, the digester volume required will be 137 000 × 30 l (30 000 × 30 gal) i.e. about 4 100 m^3 (144 700 ft^3).

A digester is usually started by mixing a large quantity of previously digested sludge ('seed sludge'), from another digester or from an open sludge digestion tank, with the raw sludge. Subsequent daily additions of raw sludge and withdrawals of digested sludge must be fairly small (generally not more than 4 per cent of the digester contents) and should be adjusted so as to maintain good gas evolution and a sludge pH value of 6·8–7·8. A method of starting up digesters which has found favour in recent years, in the case of fixed-roof digesters, is to fill the digester with settled sewage and after raising the temperature to add the 'seed' sludge followed by the gradual addition of raw sludge. This slowly replaces the settled sewage with sludge.

The reason for maintaining the pH value of the sludge in the digester between 6·8 and 7·8 is that there are at least two fermentation stages occurring and proceeding concurrently during sludge digestion, viz. an acid fermentation by acid-forming bacteria (pH 4·0–6·5) resulting in the formation of volatile organic acids of the fatty acid series, and an alkaline fermentation (pH 7·0–7·8) by methane-forming bacteria in which these acids are broken down to methane and carbon dioxide.

Examples

$$CH_3COOH = CH_4 + CO_2$$
acetic acid

$$4CH_3CH_2COOH + 2H_2O = 7CH_4 + 5CO_2$$
propionic acid

$$2CH_3CH_2CH_2COOH + 2H_2O = 5CH_4 + 3CO_2$$
n-butyric acid

It can be seen that water plays an important part in the reactions, and that the weight of gases produced can exceed the weight of sludge decomposed.

Overloading the digester by adding too much raw sludge tends to encourage undesirable acid fermentation, and gas production will then fall or even cease. Acid fermentation will also result in the production of foam on the surface of the sludge. Experience has shown that it is inadvisable to allow the volatile acids in the digesting sludge to rise above 2 000 mg/l (expressed in terms of acetic acid). If this should happen, the quantity of feed sludge added should be considerably reduced. In extreme cases addition of lime in small doses to bring the pH to 7·0 or above, or in amounts equivalent to 100–200 per cent of the volatile acid content, may be necessary (this often stops foaming). The addition of lime should only be undertaken as a last resort.

Activated and humus sludges do not easily digest alone but give good results when mixed with primary sedimentation tank sludge.

Among the many advantages of digesting sewage sludge are:

1. The volume is reduced by about two-thirds.
2. The properties of the sludge are improved, e.g. the unpleasant smell is replaced by a pleasant tarry odour, the greasy matter is very much reduced, the sludge dries more easily on beds and the product is safer to use as manure than undigested sludge which may contain pathogenic organisms.
3. The gas produced has a high calorific value (about 26 100–27 900 kJ/m^3 or 700–750 Btu/ft^3) and so can be used for power production at the works (e.g. to drive engines of the dual-fuel type). The yield of gas is good, usually about 0·31–0·62 m^3/kg (5–10 ft^3/lb) of organic matter added. Normally gas production works out at about 0·028 m^3/*capita* day (1 ft^3/*capita* day) or about 0·62–1·25 m^3/kg (10–20 ft^3/lb) of organic matter destroyed.

Assuming a detention period in the primary digestion tanks of 20–30 days, the normal loading of a digester is about 0·64–2·08 kg of organic matter/m^3 (0·04–0·13 lb/ft^3) of digester capacity/day.

The efficiency of digestion tanks can be measured by the reduction in organic matter content of the raw sludge which under favourable conditions can amount to about 50 per cent or more. A formula for calculating the amount of organic matter decomposed by digestion has been given by Van Kleeck[12]:

$$\text{Reduction in dry organic matter} = 1 - \frac{Mr \times Od}{Or \times Md} \; 100$$

where Mr = per cent mineral matter in raw sludge
Od = per cent organic matter in digested sludge
Or = per cent organic matter in raw sludge
Md = per cent mineral matter in digested sludge

Sludge Treatment and Disposal

During digestion liquor separates out, usually in bands, especially in the secondary tanks, and is withdrawn from time to time by means of specially designed valves. Since the liquor is of a very polluting character, with a high permanganate value and high content of suspended matter, it should be returned in regulated quantities to the raw sewage inlet to the works. At some works, it is first pre-treated by settlement and/or biological filtration.

A scum layer containing sludge particles, grease and oil, hair and other fibrous materials, corks, matches and other miscellaneous material always floats at the top of the sludge in a primary digester. This interferes with digestion and gas evolution. Hence excessive amounts of this scum should not be permitted to accumulate but should be controlled by recirculation using pumping, by spraying with separated liquor or by use of scum breakers.

Anaerobic bacteria are very susceptible to inhibition by toxic substances in the sludge. Thus, sludges derived from sewages containing large amounts of metallic wastes may contain high concentrations of toxic metals (zinc, copper, chromium, cadmium, nickel, etc.) which have a retarding effect upon sludge digestion and, therefore, gas production.

Many organic compounds also cause retardation of digestion. Thus it has been reported that only 0·4 mg/l of pentachlorophenol in sewage caused almost complete failure of sludge digestion[13].

Anionic synthetic detergents can also inhibit sludge digestion, even the newer 'soft' detergents. Studies on sludge digestion by workers at the Water Pollution Research Laboratory have shown that the effect is small up to about 1·5 per cent of detergents in the dry sludge solids, but above 2 per cent, the effect is more serious[14]. Some of the hard water areas in the south-east of England using large amounts of detergents have had difficulties with poorly digesting sludge for this reason. Studies at the Water Pollution Research Laboratory have shown that treatment with long-chain aliphatic amines can remedy their difficulties.

It may be useful to list some causes of digestion failure:

1. pH too high or too low.
2. Methyl orange alkalinity too low, i.e. too low buffer capacity. The methyl orange alkalinity should be about 2 000 mg/l or more.
3. Overloading of digesters.
4. Volatile acids too high. They should be kept well below 2 000 mg/l.
5. Presence of toxic substances (e.g. metals, anionic synthetic detergents, sulphate, and sulphide).

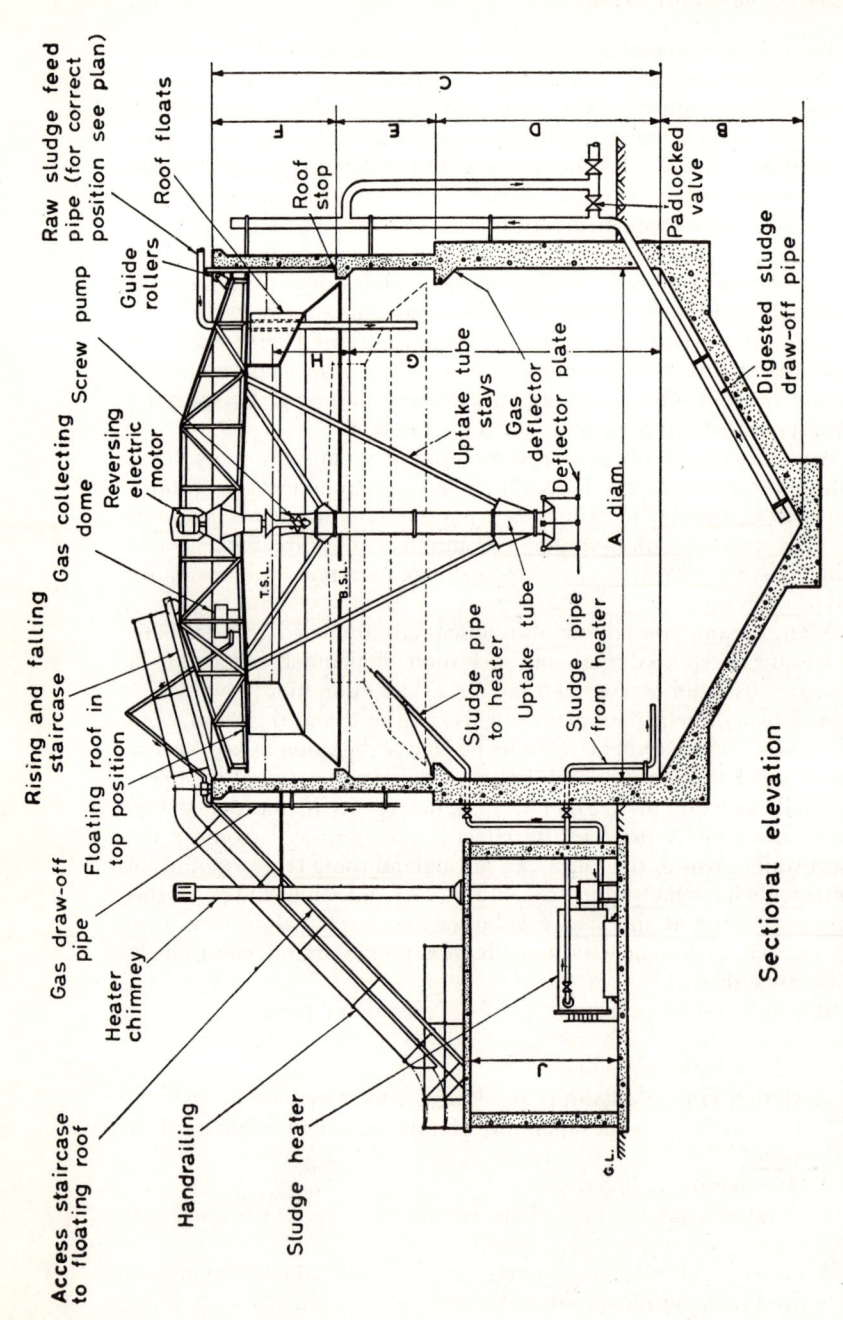

Figure 23. *Large Simplex Digestion Plant floating roof type. (By courtesy of Messrs Ames Crosta Mills & Co., Ltd., Heywood)*

6. Presence of free oxygen. Nitrate at or above 50 mg/l can cause difficulties.
7. Temperature changes.
8. Toxic ions, e.g. Na and K in large amounts. For this reason, NaOH and KOH must not be used to neutralise volatile acids in place of lime.
9. Presence of organic inhibitors, e.g. pentachlorophenol.

If gas production falls, it is advisable to first check if any leaks have occurred in the gas system before investigating other possible causes.

At nearly all the larger and medium-sized sewage works it is customary to practise sludge digestion. When the activated sludge process is used, the difficult problem of dewatering and disposing of large volumes of activated sludge is partly solved by mixing the densified activated sludge with sedimentation tank sludge and digesting the mixture.

A novel method of densifying activated sludge prior to digestion involves the use of 'flotation'[15]. This consists of injecting fine air bubbles (100 μ or less) so that the sludge floats to the top and can then be skimmed off. An activated sludge with a dry solids content of about 3 per cent can thus be obtained.

A diagram of a large Simplex digestion plant is shown in *Figure 23*.

METHODS OF DEWATERING SEWAGE SLUDGE

DRYING BEDS

The most common method of dewatering sewage sludge in this country is by means of sludge drying beds. These are generally rectangular in shape and usually have concrete floors, except in places where there is an impervious sub-soil, when the floor can be dispensed with. They are surrounded by low walls, about 0·9 m (3 ft) high, made of brickwork, reinforced concrete or even earth, but in the latter case it is advisable to cover them with asphalt or similar materials to prevent distribution of weed seeds on the sludge. There are gaps in the walls at certain points to provide access to the beds, closed by movable pieces of timber when the beds are in use. A herringbone pattern underdrainage system of field tiles is laid and channels are often provided for this purpose in concrete floors. This underdrainage system is covered by the drying bed which consists of a layer, about 300 mm (12 in) deep, of 25–38 mm (1–1½ in) clinker, on top of which is another layer, 50–75 mm (2–3 in) deep, of 6–13 mm (¼–½ in) clinker. Provision is made for removing supernatant water from the bed, often in the form of a penstock composed of pieces of timber 25–38 mm (1–1½ in) wide fitted into vertical grooves. The supernatant water is run off by removing the top pieces of timber down to just above sludge level, thus allowing the water to run into a manhole into which also flows the water from the underdrainage system.

Weir penstocks and telescopic valves can also be used for the purpose of removing supernatant water.

The liquid sludge is conveyed to the beds in pipes or channels, and distribution on to the beds is controlled by penstocks or valves. At the point of entry of the sludge the bed should be protected by means of stone flags or concrete slabs to prevent any scouring action on the clinker.

Sludge is run on a bed to a depth of about 230 mm (9 in). After a period of settlement the supernatant water is run off. The sludge is dewatered both by evaporation and by water draining through the bed into the underdrains. A number of factors affect drying bed performance, including sludge characteristics, bed design, operational techniques, whether the bed is covered or not, whether chemicals are added or not, and most important of all, weather conditions. The prolonged wet weather which is often experienced in the north of England, even in summer, causes difficulties in drying sludges. Mobile covers can now be obtained for sludge drying beds and these enable drying to be carried out even in wet weather; they work on a rail track and can easily be moved from one bed to another. Usually, however, sludge drying beds are not covered in Britain. The time taken for sludge to dry varies considerably and may be anything from a few weeks to several months. The use of covers can cut down the drying time by 25–50 per cent.

When the sludge is dry enough to be handled it is removed from the bed, a layer of the clinker about 25 mm (1 in) deep being removed at the same time; the bed has to be dressed with a fresh layer of clinker before being used again. The use of sand as a top dressing for the bed is employed at some works. Manual removal of the sludge is a time consuming and expensive operation, but the use of tractors can consolidate the clinker and thus retard drainage; there is also a risk of damage to the underdrains. These troubles are caused mainly by the tractors turning on the beds, but this has been overcome, to some extent, by constructing long comparatively narrow beds. Tractors can enter over the full width of the bed at either end and the bed is also ramped at each end to facilitate entry. Light tractors or caterpillar tracked machines are generally used, preferably with overloading actions. All these factors tend to obviate the necessity for the tractors to turn on the beds. In recent years the use of mechanical sludge

Sludge Treatment and Disposal

lifters for removing sludge from the beds has become more popular.

The area of drying beds required varies with the nature of sludge to be treated, the average local rainfall and the humidity of the atmosphere. It used to be considered adequate to provide 0·84 m² (1 yd²) of drying bed area for every 7 persons of contributory population, but experience has shown that, especially in the wetter parts of the British Isles, this is far from adequate. Experiments carried out by Swanwick and Baskerville[16] showed that 0·84 m² (1 yd²) should be provided for every 1·6 to 2·5 persons for digested sludges. When undigested sludge has to be dried on beds, it is generally necessary to provide a greater area of drying beds than for digested sludge. Where sewage contains trade wastes with high suspended solid contents this must be taken into consideration when assessing the area of sludge drying beds required.

The size of each individual bed should be based on the amount of sludge run on to the beds at one time; it is usually preferable to have a large number of smaller beds rather than a few large ones. Liquid sludge should not be run on to a bed on which there is already some partially dried sludge. Underdrainage and supernatant water from drying beds are of a grossly polluting character and should be returned to the sewage works inlet for treatment.

Sludge drying beds occupy large areas of land, and where land is not available other methods of dewatering sludge have to be employed.

Sludge dewatered on drying beds has a water content of 45–60 per cent.

A new development in drying beds is the replacement of the conventional bed by the 'Wedge Wire Filter Bed', which is made of aluminium or stainless steel wedge wire fixed together with very small spaces in between[17, 17a, 18]. To start up, the space under the wire is filled with water for 'cushioning' purposes and the liquid sludge is then allowed to run on to the top of the wire, after which the water can be slowly drained away. The period of drying sludge can thus be reduced about four times or more. The mechanical removal of the dried sludge is also simplified.

FILTER PRESSING[19]

This is an old batch process for dewatering sewage sludge, having the advantages of cheapness and of giving a product of lower moisture content than most other methods; moreover, its efficiency is independent of weather conditions. The press cake can be used as a manure on agricultural land. Automated filter presses are now available (Progress Engineers Ltd., Stoke-on-Trent, Staffs.)

The process is in use at a few works for dealing with primary sludges, especially when grease extraction is practised e.g. at Halifax, Huddersfield and Bradford. It is normally unsuitable for humus and activated sludges unless these are specially conditioned (as in the Porteous process). Other works using the process include Luton, Heckmondwike, Sheffield and Hamilton.

The sludge is generally pre-treated with lime or other suitable chemicals to assist subsequent pressing (see under chemical conditioning, p. 124).

The older type of filter press consists essentially of a series of rectangular cast-iron ridged hollow plates between which are jute or nylon filter cloths. The press is closed and the plates pressed together hydraulically or by winding screws. The sludge is forced through holes in the centre of the plates into the spaces between the filter cloths, pressure being maintained at $4 \cdot 2 – 8 \cdot 4 \times 10^5 \, N/m^2$ (60–120 lb/in^2). Water is forced through the cloth leaving behind the solids which form into a cake. Pressing time may vary from a few hours to as much as 30 h, depending on the nature of the sludge. The filter cakes, which have to be discharged by manual labour on the older types, generally have a moisture content of about 50–65 per cent.

The press liquors are highly polluting and must be returned to the sewage inlet for treatment.

At the Buxton Sewage Works where, owing to limitations of space and to high rainfall, the use of drying beds for sewage sludge dewatering is not practicable, filter pressing has been practised for well over 30 years. Copperas and lime, both available locally, are used as conditioners, and polyvinyl chloride as press cloth. The final cake has a moisture content of 65 per cent.

At Sheffield (Blackburn Meadows), there are 36 presses using a time cycle of about 4½ h and producing 30 tons/h of sludge cake with a moisture content of 55–65 per cent. The sludge (a mixture of primary and activated sludges) contains much iron from steel industry wastes so it is only necessary to add lime as conditioner. Cloths of polyvinyl chloride give satisfactory results[20].

VACUUM FILTRATION[21, 22]

Vacuum filtration of sewage sludges is the filtration of chemically conditioned sludges by applying suction or 'vacuum' for dewatering to give a cake of reasonably low water content. It is necessary to condition the sludge first by adding chemicals which cause its particles to coalesce into larger flocs.

Sludge Treatment and Disposal

Several types of filter are available for large-scale vacuum filtration of sewage sludges, e.g. Dorr-Oliver, Conkey, and Komline-Sanderson.

One, made by the Dorr-Oliver Co., consists of a hollow cylindrical segmented drum revolving slowly on a horizontal axis, partly submerged in a trough containing the conditioned sludge. The drum is covered with a suitable filter cloth (e.g. wool, nylon, glass fibre) and as it passes through the sludge, vacuum is automatically applied, causing a layer of sludge cake (usually about 3–19 mm ($\frac{1}{8}$–$\frac{3}{4}$ in) thick) to be picked up. The liquor drawn through the cloth is returned to the sewage inlet for treatment. With the emergence of the cake from the trough into the air, air is sucked through which assists in the removal of the water. Before discharge over a scraper or 'doctor', a slight air blow is automatically applied to loosen the cake from the cloth. This cycle of operations is repeated, making the process continuous. It requires some technical supervision and the cost is fairly high.

A disadvantage of the older type of vacuum filter is the occurrence of cloth blinding and the necessity of replacing the cloth fairly frequently. A new type of rotary vacuum filter, originating in the U.S.A., the Komline-Sanderson coil filter[23] uses coil springs instead of filter cloth, is self-cleansing, and can handle all types of sludge without blinding; this may replace the older type since the life of the stainless steel coil springs is said to be at least 10 years. Over 400 of these units are already in use in the U.S.A. and a number are in operation in Britain and are obtainable from the Dorr-Oliver Co. in this country.

A number of important factors affect markedly the successful operation of vacuum filtration, for example the kind and amount of chemical used; the type, physical and chemical characteristics, and nature of the sludge; the type of plant used; whether or not elutriation is used; the submergence of the drum and drum speed. The amount of coagulant used varies between 1 and 8 per cent (on dry solids basis), depending upon the coagulant, the kind of sludge and whether elutriation is used.

Table 15 shows results obtained at a few plants with various types of sludge.

In general, digested and primary sludges give cakes of much lower moisture content than activated sludges. It was found at the West Hertfordshire Main Drainage Authority's Works that nitrifying activated sludges tend to give a higher output and cakes of lower moisture content than clarifying sludges.

A new concept of 'specific resistance', i.e. the resistance of a unit weight of sludge cake per unit area at a given pressure, has been introduced by Coackley[24] and promises to be of great value in sludge filtration problems. It is given by the equation

Table 15. RESULTS OF VACUUM FILTRATION OF SLUDGES AT SEVERAL ENGLISH PLANTS

Filter used	Type of sludge	Coagulant used	Cake output kg dry solids/m² h (lb dry solids/ft² h)	Approximate moisture content of sludge cake per cent
Dorr-Oliver	Digested primary, elutriated (West Kent)	aluminium chlorohydrate	29–39 (6–8)	70–75
Dorr-Oliver	Digested primary, elutriated (West Hertfordshire Main Drainage Authority)	chlorinated copperas	about 12 (about 2·4)	70–75
Dorr-Oliver	Activated (West Hertfordshire Main Drainage Authority)	chlorinated copperas	4·4–6·8 (0·9–1·4)	84–90
Dorr-Oliver	Digested primary+activated, elutriated (Oxford)	aluminium chlorohydrate	1·9–12 (0·4–2·5)	73–80
Komline-Sanderson	Humus (Leeds)	lime and aluminium chlorohydrate	11–24 (2·2–4·9)	80
Komline-Sanderson	Primary plus humus (Scunthorpe)	lime and copperas	13 (2·7)	80

Sludge Treatment and Disposal

$$r = \frac{2bPA^2}{\eta c}$$

where A = filter area (cm^2)
P = pressure difference across sludge, sludge cake and filter cloth (g/cm^2)
η = viscosity of filtrate (poises)
c = weight of cake solids per unit of filtrate (g/cm^3)
r = specific resistance of sludge (s^2/g)

The quantity b can be calculated by a Buchner funnel test with the sludge at a standard pressure. The volumes of filtrate are recorded at known increments of time. If V is the volume (ml) at time t (sec)

$$\frac{t}{V} = bV + a$$

where a = a constant related to cloth resistance.

The slope of the straight-line graph obtained by plotting t/V and V gives b, the intercept on the t/V axis gives a.

Specific resistance is a convenient measure of the resistance of a sludge to filtration. It can be used to determine cake yields on a vacuum filter (or the time required to produce a cake in a filter press) much more accurately than was possible when using the older form of Buchner funnel test. When all the other variables are arbitrarily fixed, the cake yield on a vacuum filter has been shown by Jones and Jenkins[24a] to be inversely proportional to the square root of the specific resistance. An automatic instrument is available for the test [24b].

Coackley gives the following typical values for the specific resistance of different types of sludge:

Sludge	Specific resistance s^2/g (at pressure of 500 g/cm^2)
Primary	2·88 × 10^{10}
Digested	1·42 × 10^{10}
Activated	0·47 × 10^{10}
Digested + coagulant	0·024 × 10^{10}

It will be noticed that addition of coagulant lowers the specific resistance of sludge very considerably, i.e. makes it much easier to filter.

For satisfactory filtration, a sludge should be treated with sufficient chemical or chemicals to reduce its specific resistance to about 0·04 × 10^{10} s^2/g.

Most industrial sludges have much lower specific resistance values than sewage sludges and so are easier to dewater. Nevertheless, the amount of space for tipping dried industrial sludges is becoming more difficult to find in heavily industrialised areas.

PAXMAN ROTO-PLUG CONCENTRATOR[25, 26]

This is a new mechanical sludge-thickening device based on the American 'Roto-plug' system and manufactured under licence by Messrs. Davey, Paxman and Co., Colchester. The process involves two stages:

1. The sludge is continuously fed to a series of revolving drums rotating at about 2·8 rev/min covered with a special woven nylon fabric. Chemical conditioning is unnecessary. The liquor drains through the fabric whilst the solids are left inside the drum. The rotation causes the solids to form a cylindrical plug which becomes partly dewatered by its own weight and by the rolling action. As it lengthens, cutters cut off the ends of the rolling mass. At this stage, the moisture content of the sludge is about 85 per cent.
2. In a second stage, the thickened sludge is given additional dewatering in a special compression filter, to give a final product with a moisture content of about 70–80 per cent.

Tests at the Bletchley Sewage Works[27] showed that the capital cost of treatment is about the same as for drying beds, i.e. about £1 per head of population. Power consumption is low. The later types of Roto-plug concentrator are provided with automatic control. The method can be used for raw primary sludge, digested sludge, and activated sludge. Nevertheless, it is sometimes necessary in order to form a satisfactory plug to add fibrous matter (e.g. shredded pulped newspaper). Favourable features of the process are that it lends itself to automatic working and the machinery occupies a relatively small building.

METHODS OF DRYING SLUDGE

After sludge has been dewatered by one of the above methods it is sometimes necessary to dry it still further and convert it into a finer state of division, especially if it is to be used as fertiliser.

Air drying—The simplest method is to store the dewatered sludge in bays under cover, when further drying takes place, and then pass it through a mechanical disintegrator or grind it in a ball mill. By this method the moisture content of the sludge can be reduced to between 20 and 30 per cent.

Sludge Treatment and Disposal

Raymond system—In this system the dewatered sludge is mixed with about three times its volume of dried sludge of about 10 per cent moisture content. The mixture is processed in a cage mill through which also pass hot gases from a furnace. The sludge is broken into small particles, dried and carried into a cyclone separator by the gases. Part of the dried sludge removed from the separator is returned to be mixed with more dewatered sludge, part placed in storage containers to be used as fuel or fertiliser. The gases passing from the cyclone separator are, via a heat exchanger, conveyed back to the furnace.

Sludge dried by this method has a moisture content of 5–10 per cent. A plant of this type was installed at the Maple Lodge Works of West Hertfordshire Main Drainage Authority.

Atritor—The atritor consists of a high-speed rotor carrying square pegs which move between round pegs mounted on a stationary frame, the whole being enclosed. The device, originally used to dry and pulverise coal prior to being fed into a furnace, is successfully applied to the drying and pulverising of sewage sludge. Dewatered sludge, of about 45–50 per cent moisture content, is fed into the atritor where it is broken down in a current of hot air from a forced-draught furnace. The dried pulverised sludge passes into a cyclone separator; the steam is liberated to the atmosphere. Sludge dried by this process has a moisture content of less than 10 per cent. Atritors have been installed at Mogden, Bolton and Huyton-with-Roby.

Other types—Rotary kiln driers have been used successfully at the works of Huddersfield Corporation and West Kent Sewerage Board. A Scott drier, in which the sludge is dried on conveyor bands by hot gases in a chamber, was installed at Halifax Sewage Works.

Multiple hearth furnaces are in use in the U.S.A.; in these the sludge can be burned to a fine ash or dried to a powder of a low moisture content.

DISPOSAL AND USES OF DEWATERED AND DRIED SLUDGE

Dewatered sludge can be disposed of by tipping if suitable land is available. This should be carried out in layers which should not be too thick, otherwise the tip will heat up and spontaneous combustion may take place. Any drainage from sludge tips should not be allowed to discharge directly into a watercourse, as it is grossly polluting. Sludge can also be used to fill in depressions in land and as a surface layer on refuse tips and colliery or quarry spoil heaps and industrial waste tips. Its use on tips promotes the growth of grass, and this in conjunction with the provision of drainage tiles in the surface layer can reduce the percolation of rain water through the tip and thus reduce pollution of watercourses due to the emanation of polluting drainage from the tips. In addition it makes the tips less aesthetically objectionable.

As it has a high organic content, sludge can be burned. Its calorific value varies from less than 9 300 kJ/kg (4 000 Btu/lb) of dry solids for digested sludges to up to 20 900 kJ/kg (9 000 Btu/lb) for some primary sludges[1]. If combustion is to be self-supporting the water content of the sludge must be below 60 per cent.

There are three main types of incinerators which can be used to burn sewage sludge:

1. *Multiple Hearth*—this consists of a number of perforated hearths set one above the other. The dewatered sludge is fed in at the top and passes downwards, and the combustion gases flow in an upward direction. A raking device is provided to rake the burning sludge on the hearths so as to ensure even combustion. A plant of this type has been installed at the Sheffield Corporation's Works.
2. *Rotary Drum*—this consists of a rotating steel cylinder, with one end higher than the other. The dewatered sludge is introduced at the higher end of the cylinder and both the burning sludge and the combustion gases flow in the same direction. Both the ash and the gases are expelled from the lower end of the cylinder.
3. *Fluidised Bed*—in this type of incinerator the dewatered sludge is fed on to a bed of heated sand which is kept in a state of agitation by air blowing upwards through the bed. The ash and the gases pass into a cyclone separator.

USE AS FERTILISER

Sewage sludge, like farmyard manure, contains substances of considerable fertilizing value, such as nitrogen, phosphorus, potassium, humus, minor essential elements (especially magnesium, manganese and boron) and organic growth-producing substances. It also conditions the soil and helps to retain moisture. The nitrogen is only slowly available, and potassium is usually present in quantities insufficient to make it a balanced fertiliser. A large proportion of the valuable fertilising material in sewage, especially nitrogen, potassium and phosphorus, does not find its way into the sludge since it passes out of the sewage works in the treated effluent. Activated sludge has a higher nitrogen content than any other type of sewage sludge. Care must be exercised in using as fertilisers sludges derived from industrial sewages, as toxic metals may be present in quantities sufficient either to have adverse effects on plant and animal life in the soil, or to immobilise the phosphate as insoluble phosphate; a high grease content in the sludge will be injurious to the soil. Sludges with moisture content of 45–60 per cent taken from sludge-drying beds are not easy to spread on land and often contain weed seeds. They can also contain harmful bacteria and other micro-organisms, especially if not digested. Heat-dried sludges, being in a pulverised state, are easier to apply to land, and owing to the high temperatures to which they have been raised are usually free from weed seeds and harmful bacteria.

Nevertheless, the use of liquid digested sludges on land is gaining favour in England. It is in use for part of the sludge at the West Hertfordshire Main Drainage Authority's Sewage Works at Maple Lodge, where it is disposed of on farmland by a fleet of tanker vehicles[28]. The cost of distributing the liquid sludge in this way is very much lower than the cost of drying the sludge and then applying it to the land. Moreover, large scale trials by the Hertfordshire Institute of Agriculture showed that liquid digested sludge gave better results than did air-dried sludge[29]. Most of the fertilising value comes from the nitrogen present.

REFERENCES

1. *Ministry of Housing and Local Government. Report of an informal working party on the treatment and disposal of sewage sludge,* 1954. London, H.M.S.O.
2. ESCRITT, L. B., *Sewerage and sewage disposal: calculations, design and specifications.* 3rd Ed., 1965. London; C.R. Books Ltd.
3. IMHOFF, K., MÜLLER, W. J. and THISTLETHWAYTE, D. K. B., *Disposal of sewage and other waterborne wastes,* 2nd ed., 1971. London; Butterworths
4. JENKINS, S. H., Progress in the methods of treatment and disposal of sewage sludges, *J. Proc. Inst. Sew. Purif.* 2 (1939) 278
5. WYLIE, J. C., *Mechanised composting, Final Report, Public Works Congress and Exhibition,* 1956. pp. 856–80
6. ROGERS, C. and EVANS, G. L. K., Notes on the pressure filtration of sewage sludge at the Wandle Valley Works, *J. Proc. Inst. Sew. Purif.* 2 (1957) 137
7. CRUNDALL, S. W. F., Aluminium chlorohydrate in sewage sludge dewatering, *J. Proc. Inst. Sew. Purif.* 4 (1956) 397
8. CRUNDALL, S. W. F., MURRAY, A. and SIDLOW, R., Aluminium chlorohydrate as a conditioning agent for sewage sludges applied to drying beds, *J. Proc. Inst. Sew. Purif.* 1 (1956) 84; see also SAMBIDGE, N. E. W., *ibid* 5 (1966) 476
9. STILLINGFLEET, J. V., Filter pressing of digested sludge with chemical conditioning, *J. Proc. Inst. Sew. Purif.* 1 (1958) 41; 2 (1965) 155
10. LUMB, C., Heat treatment as an aid to sludge dewatering—ten years full-scale operation, *J. Proc. Inst. Sew. Purif.* 1 (1951) 5
11. PORTEOUS, I. K., Mechanical treatment of sewage sludge by the steam-injection method, *Munic. Engng, Lond.* 143 (1966) 2624
12. VAN KLEECK, Operation of sludge drying and sludge gas utilisation units, *Sewage Wks J.* 17 (1945) 1240
13. DOWNING, A. L. and SWANWICK, J. D., Treatment and disposal of sewage sludge, *J. Instn Munic. Engrs* 94 (1967) No. 3, 81

14. BRUCE, A. M., SWANWICK, J. D. and OWNSWORTH, R. A., Synthetic detergents and sludge digestion: some recent observations, *J. Proc. Inst. Sew. Purif.* 5 (1966) 427; see also SWANWICK, J. D. and SHURBEN, D. G., *Wat. Pollut. Control* 68 (1969) 190
15. KATZ, W. J. and GEINOPOLOS, A., Sludge thickening by dissolved air flotation, *J. Wat. Pollut. Control Fed.* 39 (1967) 946; also *Wat. Waste Treat J.* 11 (1966) 172
16. SWANWICK, J. D. and BASKERVILLE, R. C., Sludge dewatering on drying beds, *J. Proc. Inst. Sew. Purif.* 2 (1966) 153
17. SPARHAM, V. R., *The Wedge Wire Filter Bed: a review of applied and fundamental functions*, Publication No. 126, 20M, 1963. Warrington; British Wedge Wire Co. Ltd.
17a. SPARHAM, V. R. and WAIN, J. A., The development of the Wedge Wire Bed, *Wat. Pollut. Control* 66 (1967) 476
18. WILLS, R. F. and HAMLIN, M. J., Studies on the filtrability of sewage sludges, *J. Proc. Inst. Sew. Purif.* 4 (1963) 342
19. THOMPSON, J. T. and PROCTOR, J. W., Filter pressing of sludge, *J. Proc. Inst. Sew. Purif.* 1 (1938) 108
20. HOLROYD, A., The Sheffield filter pressing plant, *J. Proc. Inst. Sew. Purif.* 2 (1964) 159
21. WISHART, J. M., JEPSON, C. and KLEIN, L., Dewatering of sewage sludges by coagulation and vacuum filtration. Part I. Laboratory experiments (Buchner funnel test), *J. Proc. Inst. Sew. Purif.* 1 (1946) 110
22. WISHART, J. M., JEPSON, C. and KLEIN, L., Dewatering of sewage sludge by coagulation and vacuum filtration, Part II. Experiments with a semi-scale Dorr-Oliver vacuum filter, *J. Proc. Inst. Sew. Purif.* 1 (1947) 140
23. CLIFFORD, A. J., The dewatering of sludge by the vacuum coil filter, *J. Proc. Inst. Sew. Purif.* 6 (1962) 540
24. COACKLEY, P., Research on sewage sludge carried out in the Civil Engineering Dept. of University College London (Specific Resistance Test), *J. Proc. Inst. Sew. Purif.* 1 (1955) 59
24a. JONES, B. R. S. and JENKINS, S. H., Filtration of sewage sludge from Yardley Works (Specific Resistance and Filter leaf tests), *J. Proc. Inst. Sew. Purif.* 4 (1955) 279
24b. BASKERVILLE, R. C. and GALE, R. S., A simple automatic instrument for determining the filtrability of sewage sludges, *Wat. Pollut. Control* 67 (1968) 233
25. BERGER, O. and WARREN, A. V., The Rotoplug Concentrator; developments in design and operation, 1959–1965. *J. Proc. Inst. Sew. Purif.* 4 (1966) 327
26. SURIYADASA, R., Investigations on mechanical dewatering of digested sludge at Colchester, *J. Proc. Inst. Sew. Purif.* 4 (1966) 337; see also *Wat. Pollut. Control* 68 (1969) 224
27. TRUESDALE, G. A. and BIRKBECK, A. E., Tests of a mechanical sludge dewatering plant, *J. Proc. Inst. Sew. Purif.* 4 (1963) 307
28. KERSHAW, M. A. and WOOD, R., Sludge treatment and disposal at Maple Lodge, *J. Proc. Inst. Sew. Purif.* 1 (1966) 75
29. ANON., Liquid digested sludge on farm land an economic and nationally useful method, *Surveyor*, 125 (1965) No. 3797, 34

Chapter 9

FLOW MEASUREMENT

To enable a sewage works to be operated in an efficient manner, it is essential that on all but the smallest works adequate flow measuring facilities be provided. The usual points at which measurements should be carried out are as follows:

1. At the works inlet to measure all flows entering the works. This, besides measuring the total flows entering the works, gives an indication of the diurnal variations which take place. It also provides evidence of increases in the daily dry weather flow. Such increases can be due to the ingress of abnormal amounts of infiltration water, increases in the volume of domestic sewage or increases in the volume of trade effluent. Without proper measurement facilities such increases can gradually occur and assume considerable proportions before being detected.
2. At a point downstream from the storm sewage separation weir on the works, to indicate the volume of sewage passing forward for full treatment. Measurements taken at this point provide a check on the rate of flow at which the overflow commences to operate. They also, when considered in conjunction with the analytical results of samples taken at this point, provide evidence regarding the load being imposed on the treatment plant.
3. At a point prior to the biological oxidation units. Where units of differing types are employed, it is advisable to provide flow measurement prior to each unit.
4. Measurement of the volume of sludge being fed into sludge digestion plants, mechanical dewatering plants or other means of sludge treatment or disposal. This provides evidence regarding the amount of sludge being produced on the works and allows control to be exercised over the flow to the sludge treatment or disposal units.
5. Measurement of the flow of returned sludge and surplus sludge on activated sludge plants.

It is essential that these flow measurements be automatically recorded on a chart so that a permanent record is available, for future reference, of the daily flows and also of the fluctuations which occur. Such records are invaluable when a new sewage works or extensions to an existing works are being designed and also when trade effluent charges are being assessed. They are also useful when investigating reasons for a deterioration in the quality of the final effluent.

The usual practice, on the larger works at the present time, is to transmit the flow measurement from the point of measurement to a central panel, where all the measurements taken at the various points on the works are recorded on charts. They can thus be kept under observation by the managerial and technical staff. It is good practice to have an indicator, showing the flow, at the point of measurement, for the benefit of the plant operators. Most modern flow measurement devices are fitted with an integrator mechanism which enables the daily volume to be read off direct thus eliminating the tedious process of measuring these on the recorder charts.

The measurement of flows on sewage works falls into two main categories, namely, the measurement of flows in open channels and the measurement of flows in pipes.

MEASUREMENT OF FLOWS IN OPEN CHANNELS

Most of the usual methods employed to measure flows in open channels utilise the principle of placing a restriction in the channel and measuring the resulting head of liquid in the channel, upstream from the restriction and relating this head to the flow passing through the restriction. The restriction is usually made by constructing a weir or flume in the channel.

Weirs—Sharp-edged weirs are normally used on sewage works and thin sharp-edged metal plates are utilised. Broad-crested weirs are more suitable for measuring very large rates of flow and are more often employed in the measurement of stream flows and the overflows from impounding reservoirs.

The sharp-edged weir can take several forms, the more commonly used ones being as follows:

(a) *V-Notch Weirs*—A 90° notch is most often used and the advantages of this type of weir is that it will accurately measure small rates of flow and yet measure flows over a fairly wide range. For smaller

Flow Measurement

flows a notch with an angle less than 90° can be employed and for larger flows an angle greater than 90° can be used.

The following equation can be used to calculate the rate of discharge through a V-notch weir:

$$Q = \frac{8}{15} Cd\sqrt{(2g)} \tan\frac{\theta}{2} H^{2.5}$$

where Q = rate of discharge (ft^3/s)

g = acceleration due to gravitational force

Cd = coefficient of discharge, which is determined experimentally

θ = angle of notch

H = head of liquid over the apex of the notch (ft).

For a 90° notch, if Cd is 0·6 then:

$$Q = \frac{8}{15} \times 0·6 \times \sqrt{64} \times 1 \times H^{2.5}$$

$$Q = 2·56 \, H^{2.5}$$

(b) *Rectangular Weirs*—These are used for the measurement of the larger flows. They can be of the fully contracted type or the suppressed type.

Fully Contracted Weirs—Both the horizontal and vertical edges of this type of weir consist of sharp-edged thin metal plates and the width of the weir is less than the width of the channel in which it is constructed.

Francis's Formula for calculating the flow through this type of weir is:

$$Q = 3·33 \, (L - 0·1nH) \, H^{3/2}$$

where Q = rate of discharge (ft^3/s)

L = length of weir (ft)

H = head of liquid over weir (ft)

n = number of side contractions.

In the case of the normal single rectangular notch weir $n = 2$, but this figure will be increased in the case of a large weir which is divided into several sections across its width.

Suppressed Weirs—In this type of weir the horizontal sharp-edged plate stretches across the whole width of the channel, therefore, the number of end contractions is nil.

Francis's Formula for this type of weir is as follows:

$$Q = 3\cdot 33 \, L \, H^{3/2}$$

where Q = rate of discharge (ft^3/s)

L = length of weir (ft)

H = head of liquid over weir (ft).

There are several variations of these types of weirs such as the compound weir which is a combination of V-notch and rectangular weirs and allows a very wide range of rates of flows to be measured by one weir.

An orifice cut in a plate fixed in the channel can also be used, but

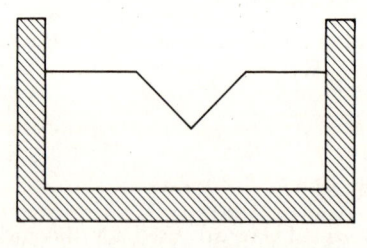

V-notch weir (90°)

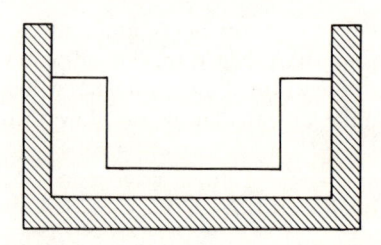

Fully contracted rectangular weir

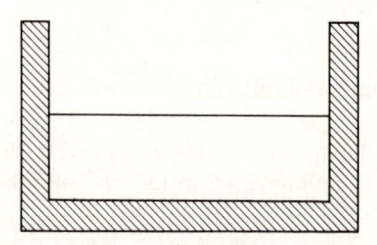

Suppressed rectangular weir

Figure 24.   Measuring Weirs

Flow Measurement 149

these are very rarely seen on sewage works due to the fact that the orifice can be restricted or blocked by rags, paper or pieces of timber, etc.

For efficient measurement it is essential to have a straight length of channel upstream from the weir, otherwise baffle plates may have to be installed in the channel.

The fall of water from the weir is known as the nappe and it is essential that the air between the nappe and the weir face be maintained at atmospheric pressure. In the case of suppressed weirs a vent pipe has to be introduced to maintain this condition.

If, as is usual, a continuous recording device is employed, the head over the weir is measured by a float situated in a chamber upstream from the weir, alongside the channel. This chamber is connected to the channel by means of a pipe, which should be of a large enough diameter to prevent clogging by sewage solids. It is advisable to provide means whereby the chamber can be flushed out. The head has to be measured upstream from the weir; measurements taken at the weir would be inaccurate because the head reduces as it gets near to the weir.

Flumes—These are usually formed by curved side contractions on the side of the channel and in some cases these are combined with a hump on the floor of the channel in the throat of the flume. The faces of the side contractions are often set at 90° to the floor of the channel, but trapezoidal or parabolical cross-sections are sometimes employed. These flumes are often called 'standing wave flumes' alluding to the standing wave which forms just downstream from the throat. The effective head is the height from the invert level in the throat to the water level in the channel upstream from the flume and, where a hump is used, from the highest point of the hump to the water level in the upstream channel. It is imperative that the downstream head does not drown the throat and thus there must be an adequate fall on the invert of the channel downstream from the throat.

A flume with vertical side contractions is the type most commonly used on sewage works. The rate of discharge can be arrived at from the following formula:

$$Q = 3 \cdot 09 \, BH^{3/2}$$

where Q = rate of discharge (ft^3/s)

B = width of the throat of the flume (ft)

H = head in the channel upstream from the flume over the invert of the throat of the flume (ft).

Measurement is usually effected, as in the case of weirs, by means of a float situated in a chamber, with a connection to the channel, upstream from the flume, the float being connected to a measuring and recording instrument. Also, as in the case of weirs, it is necessary for there to be a straight length of channel upstream from the flume.

Flumes have certain advantages over weirs in that they are not as susceptible to silting up and they create a very small loss of head.

In addition to flow measurement, flumes are often used on sewage works to control flows. They can be found in channels downstream from storm sewage separation weirs, at the outlets to grit channels and incorporated with flow distribution chambers.

MEASUREMENT OF FLOWS IN PIPES

Venturi Meters—One of the most common means of measuring the rate of flow in a pipe-line under pressure, found on a sewage works, is by using a Venturi Meter. This is effected by replacing a section of

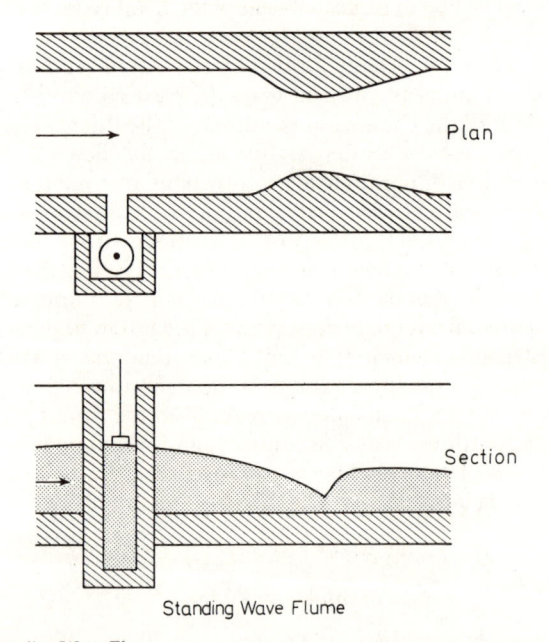

Figure 25. Standing Wave Flume

the pipe-line with one having a tapered cross section. The cross-section reduces to the throat and then increases back to the original cross-section of the pipe line (see *Figure 26*).

Flow Measurement

Because of the reduced cross-section at point Y the velocity of the flow at this point will be greater than at point X, with the corresponding result that the pressure at point Y will be less than at point X. By measuring the pressures at points X and Y these can be related to the velocity of the flow and thus to the rate of flow. Pressure tubes are

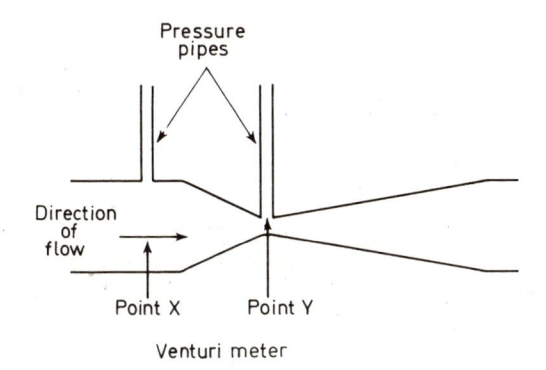

Figure 26. Venturi Meter

inserted at points X and Y and these are connected to a pressure measuring instrument which then converts these readings into rates of flow which are recorded on a chart. To prevent blockages occurring in the pressure pipes they are flushed at intervals with jets of clean water from pipes connected to them.

The rate of discharge can be calculated from the following formula:

$$Q = \frac{1}{\sqrt{\left[1 - \left(\frac{d}{D}\right)^4\right]}} \times \frac{\pi d^2}{4} \times \sqrt{(2gH)}$$

where Q = rate of discharge (ft^3/s)

D = diameter of pipe at point X (ft)

d = diameter of pipe at point Y (ft)

H = pressure head loss between points X and Y (ft)

g = acceleration due to gravitational force.

This rate of discharge is a theoretical value and a coefficient has to be applied to account for frictional resistance.

Orifice Meters—These work on a similar principle to the Venturi Meter and consist of a thin metal plate with a sharp-edged circular orifice fixed in a pipe at right angles to the direction of flow. The pressure head at both sides of the orifice is measured by means of pressure pipes which are connected to a measuring instrument. The loss of pressure head from upstream of the orifice to downstream is converted to a rate of discharge.

Magnetic Flow Meters—This type of meter has come into prominence in recent years and consists of a collar fitted round the pipe, in which a magnetic field is set up, and electrical probes inside the pipe. Its main value lies in the fact that there is no reduction in the diameter of the pipe and it is particularly applicable for the measurement of the flow of such things as sewage sludges. It is particularly important to ensure that no air pockets are formed round the probes.

MEASUREMENT OF RETENTION TIMES IN SEWAGE TREATMENT UNITS

In addition to measuring the flow through a sewage works it is often necessary and desirable to measure the actual retention time of the sewage in treatment units, as opposed to the theoretical retention time. The early methods employed entailed the use of colour tracers or chemical tracers. In the latter case a suitable chemical was introduced into the flow entering the unit and the concentration of the chemical in the flow leaving the unit was determined at known intervals of time. In the last 20 years the use of radioactive tracers for determining retention times has come into prominence and investigations into their use have been carried out by Truesdale[1] and Eden[2] amongst others. It is imperative that experiments using radioactive tracers should be carried out under the supervision of a person having had proper training in the appropriate techniques and that all necessary authorisations be obtained before the experiments are commenced.

REFERENCES

1. TRUESDALE, G., The measurement of sewage flow using Radioactive tracers, *J. Proc. Inst. Sew. Purif.* 2 (1953) 97
2. EDEN, G. E., Some uses of radio-isotopes in the study of sewage-treatment processes, *J. Proc. Inst. Sew. Purif.* 4 (1959) 522

Chapter 10

EFFECTS OF TRADE WASTES

INTRODUCTION

The most satisfactory and efficient means of disposing of liquid industrial wastes from manufacturers' premises is to discharge them into the sewers of the local authority, provided (a) the sewers and sewage works are of adequate capacity to deal with them, and (b) the waste has no damaging effects on the fabric of the sewers or sewage works or on the treatment processes. Since the Public Health (Drainage of Trade Premises) Act, 1937[1] was placed on the Statute Book the volume of trade wastes discharged into the sewers has increased considerably, and this increase is continuing. Consequently the effect of these trade wastes on the operation of sewage works has become one of the sewage works managers' major problems.

Under the Act of 1937 a manufacturer had the legal right, subject to certain conditions and safeguards, to discharge his trade wastes to the municipal sewers. The main provisions of the Act were:

(a) Manufacturers had a prescriptive right to continue discharging into the sewers any trade waste they discharged during the 12 months' period ending on the 3rd March, 1937, provided that the daily volume, rate of flow and character of the waste remained the same.

(b) Manufacturers had a right to discharge into the sewers trade wastes which were the subject of agreements made with the local authorities concerned before 1st July, 1937.

(c) Manufacturers had a right to serve a 'trade effluent notice' on a local authority for the discharge of trade waste into the sewers. This notice had to give details of the nature and composition, maximum daily volume and rate of discharge of the proposed discharge of trade waste. No trade waste could be discharged into the sewers until the expiration of two months from the day the notice was served—this was called the 'initial period'. The local authority could at any time within the initial period give a direction to the manufacturer that the trade waste could not be discharged into the sewers before a specified date. The local authority could give an unconditional consent to the discharge of the trade waste into the sewers or could impose conditions regarding the sewers into which the waste may be discharged, the nature and composition, maximum daily volume and rate of discharge, or any other matter with respect to which bye-laws could be made under this Act. In those cases where the local authority would not consent to the trade waste being discharged into the sewer or imposed conditions which the manufacturer considered unreasonable, the manufacturer had a right of appeal to the Minister of Housing and Local Government.

(d) Manufacturers could enter into agreements with local authorities for the discharge of trade wastes into the sewers. These agreements could contain clauses regulating the character, daily volume, rate of flow and temperature of the discharge, and provide for charges to be made for the treatment of the waste.

(e) Manufacturers could discharge trade wastes into the sewers subject to trade waste bye-laws made in accordance with this Act. These bye-laws could contain clauses to regulate the times of discharge, volume and rate of flow, temperature and pH of the wastes. They could also contain clauses to exclude condensing water and specified injurious constituents, and to provide for the construction of suitable inspection chambers and the installation of flow meters. These bye-laws could also provide for the making of charges for the treatment and disposal of the trade wastes. It is interesting to note that no bye-laws were made under this Act.

(f) Laundry waste was not regarded as 'trade waste' for the purposes of this Act and could be discharged to the sewers without the consent of the local authority, but bye-laws limiting the temperature and pH value could be made.

Part V of the Public Health Act, 1961[2], amended the provisions of the 1937 Act, and the main effect was to bring pre-1937 discharges under greater control. The majority of the provisions of the 1937 Act remain in force, but the principal alterations brought about by the 1961 Act are as follows:

(a) A local authority has the right to issue a direction to a manufacturer who had a prescriptive right, under the provisions of the 1937 Act, to discharge trade effluent to the sewer because he made a similar discharge in the 12 months prior to the 3rd

March 1937. This direction can contain conditions relating to the temperature, acidity and alkalinity of the effluent, and to require the manufacturer to provide sampling and inspection chambers, meters to measure the volume and rate of flow, and to keep records of the volume and rate of discharge and to make returns to the local authority concerning the volume, rate of discharge and nature and composition of the trade effluent discharged. The local authority may also make charges for the reception, treatment and disposal of the effluent. Once a direction has been issued, the local authority cannot issue a further direction regarding that discharge until two years have elapsed, unless the owner and occupier of the trade premises give a written consent to do so to the local authority.

The manufacturer has the right to appeal to the Minister of Housing and Local Government against a direction.
(b) The power of a local authority to make bye-laws relating to discharges of trade effluent has been withdrawn.
(c) The local authority has the right to include in a consent conditions, additional to those provided for in the 1937 Act, relating to the following:
1. The periods of the day during which the discharge shall take place.
2. The exclusion of condensing water.
3. The elimination or diminution of any constituent which alone or in combination with other substances in the sewer would injure or obstruct the sewers or would make the treatment and disposal of the sewage difficult or expensive.
4. The making of charges for the reception and treatment of the trade effluent. The local authority in fixing these charges must have regard to the nature and composition, volume and rate of discharge of the trade effluent, also to any additional expense incurred or likely to be incurred by accepting the trade effluent into the sewers, and to any revenue likely to be derived by the local authority from the trade effluent.
5. The provision of sampling and inspection chambers.
6. The provision of apparatus for determining the nature and composition of the trade effluent.
7. The provision of meters to measure the volume and rate of discharge of the trade effluent.
8. The keeping of records, by the manufacturer, of the volume, rate of discharge, and the nature and composition of the trade effluent, and the making of returns to the local authority giving this information.

9. The temperature of the trade effluent and its acidity and alkalinity.
(d) The drainage from premises used for agricultural or horticultural purposes, or for scientific research or experiment, is now classed as 'trade effluent'.
(e) If a local authority can satisfy the Minister of Housing and Local Government that a discharge from a laundry is likely to overload a sewer or make the treatment and disposal of the sewage difficult or expensive, or that there are any other exceptional circumstances, then the Minister can make an order to remove the exemption conferred on the discharge of laundry waste by the 1937 Act.

The main advantages of the 1937 and 1961 Acts, which apply to England and Wales only, to the sewage authorities* is that they allow them to exercise strict control over those trade effluents being discharged to the municipal sewers. Thus they can alleviate to some extent the adverse effects these effluents have on the sewerage systems and treatment works. The control of the rate of discharge is important, as trade effluents are not normally produced from the manufacturers' processes at a uniform rate, but in flushes, and generally only during the day-time working hours. This results in peak flows at the sewage works coinciding with the peak strengths of the sewage, with consequent great variations in the load on the treatment plant. The provision of a balancing tank at the manufacturers' premises often has a beneficial effect because in addition to providing a uniform rate of flow into the sewers over any desired number of hours in the day, it also helps to balance out the strength of the waste liquors. To conform with the conditions laid down in agreements, consents and directions pre-treatment has in many cases to be given to the waste liquors on the manufacturers' premises before they are discharged into the sewers.

CHARACTERISTICS OF TRADE WASTES

The main characteristics of trade wastes which have adverse effects on sewerage systems and treatment processes are given below. It should be remembered, however, that trade wastes vary greatly in nature and composition, and the waste from an individual factory may have only one or a number of these characteristics.

* In London, although these Acts broadly apply, they are subject to many modifications, amendments and exceptions.

HIGH SUSPENDED SOLIDS CONTENT

Trade wastes having a high suspended solids content can cause silting and blockages in sewerage systems, especially if the solids are of a fibrous character, such as hair, wool and cotton waste. They also cause an increase in the amount of screenings and sludge to be dealt with at the sewage works, thus raising running costs. If the suspended solids are of such a nature that they are not easily broken down by anaerobic bacteriological action, they can cause problems at the sewage works, as these solids can take up capacity in the sludge digestion tanks without any corresponding increase in gas production, thus creating difficulties in maintaining the temperature of the contents of the tanks. Cellulose fibre is typical of this class of solid matter.

Screening and sedimentation in tanks before discharge to the sewers will alleviate the problem and, if the tanks are of the fill-and-draw type, balancing of the flow and strength of the liquor can also be achieved.

HIGH OXYGEN DEMAND

Trade wastes which have a high B.O.D. or permanganate value can place a heavy load on the biological oxidation processes at a sewage works. High oxygen demand is usually due to the presence of organic matter or inorganic reducing agents in the waste liquors. Some wastes have a high permanganate value with a comparatively low B.O.D. This can be due to the presence of substances which are resistant to biological oxidation or to the presence of substances which inhibit biological activity. A comparison of the B.O.D. and permanganate value can often give some indication of the treatability of a trade waste.

Pre-treatment consisting of sedimentation aided by mechanical flocculation and chemical coagulation can in some cases reduce the oxygen demand. Towers filled with plastics filter media, on to which the waste liquors are dosed at a high rate, can also be used to effect a considerable reduction in the oxygen demand before the liquors are discharged into the sewers. In recent years much work has been carried out on treating trade wastes in activated sludge plants and very promising results have been obtained in the reduction of the oxygen demand. In a number of cases nutrient salts have to be added to the waste liquors to promote bacterial growth.

HIGH ALKALINITY OR ACIDITY

Waste liquors which have a high acidity or alkalinity can have an injurious effect on the fabric of sewers and sewage works. They may likewise have inhibitory effects on the biological oxidation processes, especially if causing great variations in the pH of the sewage entering the sewage works. High alkalinity can also retard sedimentation in the primary sedimentation tanks. It is desirable that trade wastes discharging into sewers should have a pH value at least within the range 5·0–11·0, but where the waste has to be treated at small sewage works narrower limits may have to be set. In some factories both acidic and alkaline waste liquors are produced, and by judicious mixing, an effluent having a pH value within the above range can often be produced. Where the waste liquors are predominantly alkaline or acidic, neutralisation has to be resorted to. Flue gas (with a carbon dioxide content of 12–18 per cent) is sometimes used to reduce the alkalinity of waste liquors. The liquor has to be brought into good interfacial contact with the flue gas by means of sprays, scrubbing towers, diffusion of the gas through, or surface agitation of the liquor. It must be noted, however that the use of flue gas obtained from the combustion of coal might well mean that scrubbing devices will have to be installed to remove phenols, etc.

HIGH TEMPERATURE

Discharges of hot liquors can constitute a danger to men working in the sewers and may also produce objectionable odours causing a nuisance to the public. Moreover, high temperatures increase the corrosive action on the sewer fabric. The temperature of waste liquors discharged into the sewers should not exceed 43°C and liquors of higher temperature should be cooled before discharge; with very hot liquors a heat recovery plant may be a practicable proposition. A watch should always be kept on the possibility of boiler blow-down water being discharged into the sewers.

INFLAMMABLE LIQUIDS

Volatile liquids having a low flash-point should not be discharged into sewers, as they can cause fires and explosions. Examples of such liquids are petrol, benzene, carbon bisulphide, acetone, methylated spirits, white spirits and paraffin. Drainage from impervious areas in garages, vehicle repair shops and washing bays, should be

Effects of Trade Wastes

passed through an oil and petrol trap before being discharged into a public sewer.

HIGH OIL AND GREASE CONTENT

Waste liquors having a high oil or grease content can cause trouble both in the sewers and at the sewage works. In the sewers the oil or grease can combine with solid matter to cause an impediment to the flow, in serious cases even a complete blockage of the sewer. At the sewage works oil and grease can lead to clogging of screens and percolating filters and also reduce the efficiency of an activated sludge plant by forming a large quantity of grease balls.

A large proportion of oil and grease can be removed from trade wastes before discharge by the provision of a suitable oil trap. Where, however, the oil is in an emulsified state, e.g. cutting oil, the liquor has to be treated with chemicals such as aluminium sulphate before being passed through the oil trap. The same applies to wool-scouring wastes in which the grease is usually present in a saponified state; the liquor has to be treated with acid or alumino-ferric, after which the grease can be removed in sedimentation tanks. In some areas of the West Riding of Yorkshire where a large amount of wool scouring is carried out, the waste liquors are allowed to be discharged into the sewers untreated and the grease is removed at the sewage works.

INJURIOUS AND INHIBITORY CONSTITUENTS

Some trade wastes contain certain substances which can cause injury to personnel and to the fabric of sewers and treatment works and which inhibit biological activity in the treatment processes. Some of the types of substances in this category are

(a) *Sulphides*—These can give rise to serious troubles. They can cause corrosion of the sewers especially at the soffit and at the waterline. If hydrogen sulphide is liberated it can have a serious effect on men working in the sewers; fatal accidents have been known to occur. Sulphides have a high permanganate value and can also have a toxic effect on bacterial life in the treatment processes. For these reasons some local authorities prohibit the discharge into the sewers of trade wastes containing sulphides, while others strictly define the permissible sulphide content of discharges. A limit of 15 mg/l in the total flow in a sewer has been suggested in the past, but by modern standards this is rather high. It is desirable to limit the amount of sulphide in trade effluent discharges to sewers to 5 mg/l or less, wherever possible.

The sulphide content of trade wastes can be reduced or removed by chlorination, precipitation as ferrous sulphide with copperas and lime, or by liberation of the sulphide as hydrogen sulphide by acidification and air-blowing where, however, steps have to be taken to deal with the poisonous hydrogen sulphide evolved. The sulphide content of trade wastes has in certain cases been reduced by treating them on percolating filters prior to discharge into the sewers. A method of sulphide removal which has been developed in the leather industry[3] consists of aerating the liquor in the presence of a manganese salt (chloride or sulphate) which acts as a catalyst.

(b) *Cyanides*—The presence of cyanides in sewage can be dangerous to men working in sewers, especially if the sewage becomes acidic, when hydrogen cyanide will be liberated. If the concentration in sewage exceeds 2 mg/l (as CN) the performance of activated sludge plants will be affected, but percolating filters can gradually adjust themselves to concentrations of up to 30–40 mg/l so that eventually there is no serious deterioration in the quality of the effluent from the sewage works. Research work at the Water Pollution Research Laboratory[4,5,6] has shown that cyanides of cadmium, zinc and potassium of a concentration of at least 100 mg/l can be completely broken down to oxidised and ammoniacal nitrogen on percolating filters, without addition of sewage, if the filters are gradually acclimatised to the cyanide.

Excessive concentrations of cyanide can result in cyanides being present in the sewage works effluent, which could have serious effects in the watercourse into which the effluent is discharged. Their presence can temporarily reduce the amount of gas produced in a sludge digestion tank.

Cyanides can be removed from trade wastes by precipitation with copperas and lime, or by liberating them as hydrogen cyanide by means of acidification and air-blowing. Great care has to be taken in the latter case to deal satisfactorily with the extremely poisonous hydrogen cyanide and it is rather a dangerous process to adopt. Chlorination can be used to treat trade wastes containing cyanides and convert them to cyanates. The pH value should, however, be 10·0–11·0 or higher because at lower pH values poisonous cyanogen chloride (CNCl) can be formed and evolved as a vapour. The presence of nickelocyanides results in the reaction proceeding very slowly, due to the stability of these compounds and a great excess of chlorine is required.

Ozone can be used to oxidise cyanides, the reaction proceeding according to the equation:

Effects of Trade Wastes

$$CN' + O_3 = CNO' + O_2$$
$$2CNO' + 3O_3 + H_2O = 2HCO'_3 + N_2 + 3O_2$$

(c) *Phenols* have a toxic effect on living organisms and can reduce the activity of activated sludge plants, percolating filters and sludge digestion processes, but under suitable conditions can be broken down by biological oxidation.

(d) *Formaldehyde* is a reducing agent and strongly bactericidal, but Waldmeyer[7] showed that trade wastes containing appreciable amounts of formaldehyde could be treated in an activated sludge plant. Over 70 per cent purification was achieved in terms of the 4 h permanganate value, provided the wastes contained at least 5 per cent of domestic crude sewage.

(e) *Toxic metals*—Certain metals have a toxic effect on the biological activity in the oxidation and sludge digestion processes at the sewage works. These include chromium, nickel, lead, copper, zinc, silver and mercury, and their presence in trade wastes should be strictly controlled. Fortunately, many of these metals are partly precipitated and settle out with the sludge in the primary sedimentation tanks, thus reducing their effect on the biological oxidation processes.

Chromium has a strong toxic effect on biological activity, especially when present as chromate or dichromate. Burgess[8] stated that 1 mg/l of chromium, as Cr^{VI}, will impair the activity of percolating filters and 2 mg/l that of activated sludge plants. If the amount of soluble chromium salts present in sludge is greater than 1 mg/l the rate of digestion will be reduced, and 200 mg/l of trivalent chromium will have the same effect.

Chromium in its trivalent form (i.e. as chromic salt) can be removed from trade wastes by precipitation with an alkali such as lime, but not so in its hexavalent form (e.g. as chromate) unless reduced to the trivalent salt. This can be brought about be reducing agents such as copperas, sodium metabisulphite or sulphur dioxide, followed by treatment with lime to precipitate the chromium in the form of its hydroxide. When copperas is used the pH of the liquor must be brought to below 3·0 with sulphuric acid to obtain satisfactory reduction, but with sulphur dioxide the reduction takes place satisfactorily within the pH range 1·0–5·5. Large quantities of sludge containing ferric hydroxide are produced when copperas is used, and it is not easty to recover chromic oxide from such a mixture. Chromium present in its hexavalent form can be precipitated from trade wastes as the sparingly soluble barium chromate by treatment with barium chloride and sodium carbonate.

Nickel—A concentration of 1–3 mg/l of nickel in sewage can have adverse effects on percolating filters and activated sludge plants.

In sewage sludges, where the concentrations are usually greater, it will retard anaerobic digestion.

Zinc—If the concentration of zinc in sewage exceeds 5 mg/l it can cause deterioration in the biological oxidation processes, but in sewage sludge it has to be present in much greater concentration before having any serious effect on the anaerobic digestion processes. It should be remembered that small amounts of zinc are present in many domestic sewages. The amount of zinc in trade wastes can be reduced by precipitation as an hydroxide by the addition of lime, but the sludge produced is difficult to dewater.

Copper—A large proportion of any copper present in sewage is removed in the primary sedimentation tanks, but the aerobic biological processes can only tolerate a concentration of 1–3 mg/l. The presence of copper in trade wastes discharging into sewers results in a build-up of the copper concentration in the sludge removed from the primary sedimentation tanks, and this can have a serious effect on sludge digestion processes. It has been found[9] that while a concentration of copper in the sludge of 2 500 mg/l (dry basis) had only a negligible effect upon gas production, a concentration of 13 000 mg/l resulted in only 35 per cent of the normal volume of gas production.

Copper can be removed from trade wastes by precipitation as its hydroxide by the addition of lime, by electroysis—the metallic copper being deposited on the cathode—or by passing the liquor through a tank containing scrap iron; the iron passes into solution and a sludge containing copper is deposited in the tank from which the copper can be recovered.

Lead—An appreciable proportion of the lead in sewage is removed in the primary sedimentation processes, as many lead salts are virtually insoluble in water.

Silver and mercury—These metals have a toxic effect on the organisms found in biological oxidation and sludge digestion processes.

Owing to their value these metals are rarely found in sewages in harmful concentrations.

Other substances, such as arsenic, iron and perborates, are also toxic to living organisms if present in sufficient concentration. Some organic compounds, which can be present in trade wastes, can be particularly harmful to sewage works processes, e.g. pentachlorophenol can seriously inhibit sludge digestion and allyl thiourea is a powerful inhibitor of nitrifying organisms.

SYNTHETIC DETERGENTS

Synthetic detergents are increasingly used in manufacturing processes, with the result that many trade waste discharges contain appreciable amounts of detergents. This increase is mainly due to the fact that, unlike soap, they are unaffected by hard water; hence they are particularly attractive in hard water areas. They can also be used in acid solutions.

The presence of synthetic detergents in sewage can produce troublesome effects at the sewage works. On the surface of activated sludge plants high banks of foam are produced which become airborne in windy weather and create a nuisance. At certain sewage works there has been much evidence of greater difficulty in treating sewage containing detergents and, in the woollen districts of Yorkshire, in the recovery of wool grease from sewage[10, 11]. Hurley[12, 13] showed that Teepol (a sodium higher alkyl sulphate) did not materially affect the oxidation of sewage and, being almost completely removed from sewage by biological filtration, did not normally cause foaming in streams. On the other hand, Lumb[14] proved that small concentrations (20 mg/l in crude sewage) of detergents of the alkyl aryl sulphonate class, as well as non-ionic detergents do, in fact, adversely influence the oxidation of sewage by percolating filters and the activated sludge process; they are incompletely oxidised biologically, and final effluents, therefore, could contain enough detergents to give rise to foam on agitation.

The changes which have taken place in the types of detergents manufactured and used, and their effects on sewage treatment, are dealt with in Chapter 12.

The sewage works manager has little control over the amount of detergent in domestic sewage. Strict attention should therefore be paid to that of trade waste discharges to ensure that the total amount present in the sewage does not get beyond reasonable limits, otherwise trouble can be caused both at the sewage works and in the rivers into which their effluent discharges.

RADIOACTIVITY[15, 16]

Waste waters containing radioactive isotopes can arise from Atomic Energy Authority premises, nuclear power stations and hospitals, universities and research organisations. Experimental workers at some sewage works also make use of radioactive isotopes (see Chapter 9). Of the three kinds of radioactive isotopes (α-, β- and γ- emitters), the β-emitters (e.g. P^{32}, Na^{24}, I^{131} and Au^{198} occurring in hospital effluents) are the most likely to be found in sewage. Some isotopes emit both β and γ radiation (e.g. Br^{82}). Ra^{226} (a naturally occurring isotope) emits all three types of radiation and is the most dangerous of the α emitters. Sr^{90} is the most hazardous β emitter.

Under the Radioactive Substances Act, 1960 which applies to Great Britain and Northern Ireland, the amount of radioactivity in sewage and trade wastes is controlled by a Government Radiochemical Inspectorate—*not* by River Authorities or other local authorities. Persons holding radioactive materials* must register with the appropriate Ministry (the Ministry of Housing and Local Government in England and Wales, the Secretary of State in Scotland, and the Ministry of Health and Local Government in Northern Ireland). The disposal of any radioactive wastes to sewers, rivers, estuaries, tips, or underground must be by recognised methods and is permitted only after a licence authorising such disposal has been granted.

The following upper limits have been suggested for this country[15]

	Not more than $\mu c/ml$[†]	pc/l[†]
Momentary quantity of radioactivity in discharge (workmen present in sewers)	0·1	10^8
Sewage from hospitals and other establishments (average concentration in sewer)	10^{-4}	10^5
Direct discharge to rivers	10^{-5}	10^4

These limits are only intended to serve as rough guides and are subject to alteration, depending on the nature of the isotopes present and upon local factors, such as the flow in the river and the use to which the river is applied. Most natural waters have very slight radioactivity, usually within the range 1–1 000 pc/l.

In general, if an effluent's radioactive content is below the lowest permissible amount of radioactive isotope or isotopes recommended by the International Commission on Radiological Protection[1], it should be safe to discharge it to a watercourse used for drinking purposes or for watering cattle. The figure for Sr^{90}, the most dangerous β emitter, is 10^{-6} $\mu c/ml$ (1 000 pc/l), lower than that of any other β emitter.

---

* Atomic Energy Authority premises and nuclear power stations are already covered by special legislation and are subject to Government control.

† *A curie* is the quantity of any radioactive isotope which disintegrates at a rate of 3.7×10^{10} disintegrations per second. It was originally intended to be the number of disintegrations per second occuring in 1 gramme of radium.

It is a rather large unit so smaller sub-multiples are mostly used.

1 microcurie (μc) = amount of radio-isotope disintegrating at rate of 3.7×10^4 disintegrations per second.

1 picocurie (pc) = a millionth part of a microcurie and was formerly known as a micromicrocurie. Its use has the advantage of giving a series of whole numbers and so avoids the use of negative exponents.

Since bacteria and other low forms of life are far more resistant to the effects of radiation than are human beings and animals, the treating of sewage by activated sludge and percolating filters is unlikely at present to be affected by the presence of radioactive isotopes.

Methods of treatment of radioactive wastes, before disposal, include storage (to bring about decay of radioactivity), but this is suitable only for isotopes with a low half-life, evaportion, fixation in a solid media to give a glass which can be buried or disposed of at sea, ion-exchange, chemical precipitation, and biological methods. In fact, the methods are similar to those used for other trade wastes, but, of course, much more elaborate precautions must be taken to ensure that the processes are adequate, foolproof and free from hazard.

TYPES OF TRADE WASTES

A diverse variety of waste liquors can be produced from the wide range of industrial processes. *Table 16* gives characteristic figures of trade waste samples. Some of the main types of processes which produce appreciable quantities of waste liquors are discussed below.

COTTON BLEACHING, DYEING, PRINTING AND FINISHING[17a]

Bleaching—(a) *Kier liquors.* In the traditional process cotton yarn or pieces are boiled in a solution of caustic soda, soda ash or lime, usually under pressure, in large vessels known as kiers. The waste liquors from this process are known as 'kier liquors'. They have a high alkalinity and high permanganate value but their B.O.D.:4 h permanganate value ratio is usually lower than that of domestic sewage due to the presence of substances which are not amenable to biological oxidation. However, these liquors comprise only a comparatively small portion of the volume of waste liquors produced in cotton bleaching. As these liquors are usually run off in flushes, it is advantageous to provide a balancing tank so that they can be discharged at a uniform rate and be diluted with the weaker wash waters before entering the sewer. It is sometimes necessary to neutralise kier liquors before discharge to a sewer, and this often results in coagulation, with a consequent reduction in the permanganate value of the supernatant liquor after sedimentation. Some research work has been carried out on the anaerobic digestion of kier liquors[18] in an effort to reduce their strength, but the results were not particularly good.

(b) *Wash waters.* These comprise the greater part of the volume of waste liquors produced in cotton bleaching. They are fairly weak—virtually weak solutions of kier liquor, hypochlorite, acid, soap and wetting agents—and often contain an appreciable amount of suspended solids, much of which consists of cotton lint. It is advisable that these liquors be passed through screens or sedimentation tanks before being discharged into the sewers.

Table 16. ANALYTICAL FIGURES OF SOME TRADE WASTE SAMPLES

Type of waste	pH	4 h Permanganate Value mg/l	Suspended solids mg/l	Other determinations mg/l
Straw kier liquor	14·2	74 000	4 782	
Sulphite cellulose (wood pulp) liquor	3·4	58 000	104	
Crude gas liquor	8·8	20 400	21	phenols 12 500 amm. N 14 000 alb. N 260
Spent gas liqor	8·6	11 800	402	phenols 2 500 amm. N 2 600 alb. N 200
Brewery waste	5·2	2 080	779	B.O.D. 4 550
Cotton bleaching kier liquor	13·2	4 640	11	
Sulphide dyeing waste	10·6	420	93	sulphide (as H_2S) 32·9
Towel bleaching wash waters	3·9	12·8	105	free chlorine 8·9
Calico printing waste	11·2	400	132	
Woollen piece goods scouring waste	11·0	908	195	
Electroplating waste	1·4	6·6	24	chromate (as Cr) 14 copper 15 nickel 22·5 cyanide (as CN) 8

(c) *Waste hypochlorite and acid*—The amount of waste hypochlorite and acid to be disposed of is fairly small in volume, but these liquors should be run off into the main flow of wash waters slowly and not simultaneously, in order to prevent any evolution of chlorine.

Dyeing—A great variety of dyes are used, including azoic, direct, vat and sulphide dyes: spent liquors from sulphide and vat dyeing are the strongest. Spent vat dyes are very alkaline and have a fairly high permanganate value; in some cases their alkalinity may have to be reduced with acid before discharge into the sewers. Spent sulphide liquors are dark-coloured, highly alkaline, foul liquors with high sulphide content and high pH and permanganate values. Neutralisation of these liquors with acid has certain difficulties, as owing to local excesses of acid, hydrogen sulphide may be liberated. Sulphides can be removed by oxidation with chlorine or hypochlorite, or by precipitation with copperas, but if in the latter case the liquor is strongly alkaline, the pH value will have to be reduced to about 8·6 to obtain satisfactory precipitation.

Printing—Machine printing gives rise to large volumes of waste liquors, produced by washing of the cloth after printing and by washing of the printing machines and colour mixing vessels. These liquors have a fairly high permanganate value owing to the presence of starches, gums and resins used as vehicles for the colour. With the newer types of colours, volatile liquids such as white spirit are sometimes present in the waste liquors. There has been a reduction in the amount of machine printing carried out in Great Britain in recent years, but a considerable amount of block and screen printing is done. The latter processes produce waste liquors similar to those from machine printing, but they are usually smaller in volume.

Finishing—The main constituent of wastes from this section of the industry is starch, which has a high oxygen demand and easily putrefies. As much as possible of the starch in solid or semi-solid form should be separated and collected and be disposed of by burning or burying.

Mercerising—This consists of treating the cotton, either in yarn or piece form, with a strong caustic soda solution, after which it is washed with water and a weak solution of acid. The waste liquors are predominantly alkaline, but the alkalinity can be reduced by adopting a counter-flow system of cotton washing and by re-using the stronger caustic soda liquors after evaporation in a recovery plant. In certain cases the waste liquors may have to be neutralised with acid before being discharged into the sewer.

In recent years a great deal of work has been carried out on the biological oxidation of waste liquors from the cotton bleaching, dyeing, printing and finishing industry, on the manufacturers' premises.

Activated sludge plants including the Pasveer ditch, and filters including those constructed with plastics media, have been used and encouraging results have been obtained. The control of the pH value of the feed to the oxidation plant has been found to be important especially where great fluctuations occur, and nutrient salts have been used in some cases with success in order to encourage bacterial action.

CARBONISATION OF COAL

When coal is distilled in retorts to produce coal gas and coke, strong waste liquors are produced which come from a number of stages in the process:
- (a) Retort house liquor which condenses in the main through which the gas passes after leaving the retort.
- (b) Condenser liquor from the water-cooled condensers through which the gas passes.
- (c) Scrubber liquor, the water used to scrub the gas to remove ammonia.
- (d) Drainage from purifiers.

All these liquors are run into a well, the tar layer which separates out being removed. The aqueous layer, known as 'crude ammoniacal liquor', contains much ammonia (both free and combined), pyridine and other organic bases, monohydric and polyhydric phenols, cyanides, sulphides and thiocyanates. It is a strong, foul liquor and can have a 4 h permanganate value as high as 20 000 mg/l. This places a great load on the sewage works if the liquor is discharged into the sewers, and the high ammonia content together with the presence of sulphides, cyanides and phenols can create difficulties in the biological treatment processes.

The amount of crude ammoniacal liquor produced varies in volume, but the average is about 147 l/tonne (33 gal/ton) of coal carbonised in horizontal, and about 228 l/tonne (51 gal/ton) in vertical retorts[19].

The only satisfactory method of dealing with this waste liquor is by biological oxidation at a sewage works in admixture with domestic sewage. 'Crude ammoniacal liquor' should not be discharged into the sewers without some pre-treatment, but where this is unavoidable the volume should not exceed 0·25 per cent of the total volume of sewage to be treated at the sewage works[20]. This usual method of pre-treatment prior to discharge is by distilling this type of liquor in a current of steam, in a 'Concentrated Ammoniacal Liquor Plant'. Ammonia, hydrogen sulphide and hydrogen cyanide are driven off, and the liquor left is known as 'spent liquor'. This, although still strong and foul, is weaker than 'crude ammoniacal liquor', has a lower permanganate value and contains less ammonia and very little sulphide and cyanide. The volume of the 'spent' liquor is greater than the volume prior to distillation owing to the condensation of steam. 'Spent' liquor should not exceed 0·5 per cent of the total volume of sewage to be treated at a sewage works[20].

The waste liquors produced during the manufacture of metallurgical coke in coke ovens are similar in character to those from gas works but are usually weaker, they contain less polyhydric phenols and thiocyanates but usually have a higher cyanide content.

Much research work has been done on the treatment of neat coke oven effluents and gas liquor by biological and other methods. With the weaker coke oven effluents some success has been achieved by the use of the activated sludge process, and a plant using this method has been working for some years in South Wales at the Maritime Coke Works.

Other waste liquors produced at a gas works include cooling and coke-quenching water, gas holder overflow water and waste liquors from producer and water gas plants. These are much weaker than 'crude ammoniacal liquor' and easier to deal with, although steps may have to be taken to remove small particles of coke from the coke-quenching water before discharge into the sewers. Difficulties can be experienced, however, when all the water from a gas holder has to be discharged (e.g. during repairs), as it is usually large in volume and can also cause noxious smells to arise from the sewers. Fortunately complete run-off of a gas holder is a rare occurrence.

The steam-assisted distillation of the tar from gas works produces a very polluting waste liquor which has a high phenol content (up to 20 000 mg/l—mainly monohydric phenol) and a high 4 h permanganate value (about 50 000 mg/l) and contains sulphides and ammonia.

The development of natural gas resources and the production of gas from oil is resulting in a reduction in the amount of gas being produced from coal in Great Britain. There will, therefore, be a consequential reduction in the amount of gas liquor to be treated at sewage works.

PULP, PAPER AND BOARD MANUFACTURE

Paper and board (which is a rough thick paper) are made from pulp composed of cellulose fibres, re-pulped waste paper or a mixture of both.

Pulp manufacture—There are three chief methods of pulp manufacture.

(a) Mechanical grinding of wood to produce 'mechanical wood pulp'.
(b) Digesting wood with a sulphite solution. The waste digestion liquors are very strong, being acidic and having a very high 4 h permanganate value which can be as much as 70 000 mg/l. Very few firms manufacture pulp in this manner in the British Isles, and evaporation is the only satisfactory means of dealing with this liquor. The concentrated liquor has a number of commercial applications, such as use as a road material and for tanning purposes. The stronger wash waters can also be evaporated, but large volumes of weaker wash waters which still contain much organic matter, and of waste bleaching liquors are also produced.
(c) Digesting straw, rags, esparto grass, cotton linters, cotton waste, jute and hemp with caustic soda in kiers. The waste kier liquor is very alkaline and has a high 4 h permanganate value. The strongest liquor, from the digestion of straw, can have a 4 h permanganate value as high as 74 000 mg/l. The kier liquors often contain substances such as lignin which are resistant to biological oxidation, and when treated at sewage works can result in the final effluent having a higher than normal 'residual' permanganate value. Some of the liquors, especially those from the production of high-strength fibres, contain sulphides. Where the volume of kier liquor is small and can be treated with very large volumes of domestic sewage, it is sometimes possible to discharge it at a uniform rate into the sewers after neutralisation. In some cases, however, it has to be dealt with in a caustic soda recovery plant in which it is evaporated, then incinerated to remove the organic matter, and finally re-causticised with lime to recover the caustic soda. An alternative method is to reduce the volume by evaporation and then to dispose of the concentrated liquor by disposal at sea. The strong first washings can also be treated in a caustic soda recovery plant, or by evaporation and sea disposal, but the large volumes of the weaker wash waters are usually too weak for recovery of caustic soda to be a sound economic proposition and the large volumes make evaporation too costly. These liquors, together with the waste liquors from pulp bleaching, can be discharged into the sewers when there is sufficient capacity available in the sewers and at the sewage works, provided steps are taken to remove any excessive amount of fibre prior to discharge.

Paper and board manufacture—The pulp or re-pulped waste paper is mixed with water, and the fibres are broken up to the required length in 'beaters'. This beating process can add to the B.O.D. of the waste liquors. Aluminium sulphate and adhesive agents, such as size or resin, are added, together with fillers such as china clay and any dyestuff which may be necessary. The pulp is then run on to an endless band of wire gauze through which the water passes, leaving behind a sheet of wet paper which is dried on hot rollers.

The wastes from paper and board making consist mainly of (a) liquors from the preparation of the fibre before it is fed on to the paper-making machine; (b) 'back water', i.e. the water collected under the paper-making machine; and (c) water used for washing down machines and floors. These waste liquors contain appreciable quantities of fibre and fillers and are in some cases acidic in nature. Owing to the poor quality of the raw material used, the waste liquors from the manufacture of board usually have a higher permanganate value than those from the manufacture of paper. As the waste waters contain a considerable amount of suspended solid matter, it is desirable to remove as much of this as possible before the liquors are discharged. This is particularly important if the liquors are to be treated at a sewage works utilising sludge digestion, as the cellulose fibres would take up capacity in the digestion tanks without producing a comparable amount of gas. Valuable raw material can be recovered by re-using as much 'back water' as possible in the manufacturing processes or by the abstraction of fibre from the waste liquors by means of saveall or by flotation. Synthetic plastic media have been used for the biological treatment of paper mill wastes[20a].

VISCOSE RAYON MANUFACTURE

Viscose rayon is made from cellulose pulp manufactured from wood or cotton linters. The pulp is allowed to stand in a solution of caustic soda and, after pressing to remove the alkaline liquor, is broken up, mixed with carbon disulphide and dissolved in a weak solution of caustic soda. This is followed by extrusion through fine jets into a bath of dilute sulphuric acid containing zinc sulphate. The filament is washed with water, treated with sodium sulphide to remove the precipitated sulphur, then washed again and, in some cases, bleached.

The waste liquors from these processes are:

(a) Alkaline liquors from the initial treatment of the cellulose
(b) Alkaline liquors containing polysulphides
(c) Acidic liquors containing sulphates and zinc
(d) Bleach liquors and wash waters

Most of these liquors contain viscose and hemicelluloses.

Liquors (a) and (c) can be used to neutralise each other, but before discharge into the sewers, care must be taken that the amount of zinc present is not sufficient to have an adverse effect on the biological treatment processes at the sewage works. The alkaline liquors containing sulphides can cause trouble in the sewers and at the sewage works, but Hughes[21] has described how such liquors can be treated on percolating filters with a dosage of 300–360 l/m^3 (50–60 gal/yd^3) of medium per day to produce an effluent having a sulphide content of about 1 mg/l, for discharge into the sewers.

WOOLLEN MANUFACTURE

Raw wool contains impurities such as natural grease and vegetable and inorganic debris which have to be removed by washing and scouring in a soda ash and soap solution. The resulting waste liquor is alkaline, greasy and highly putrescible, and has a very high 4 h permanganate value. If these liquors are discharged into the sewers, the high grease content can have a serious effect on the treatment processes at the sewage works, quite apart from placing a heavy organic load on the works. Therefore steps have to be taken to remove as much of the grease as possible, either prior to discharge or at the sewage works itself. The grease is present in the liquors in the form of salts of fatty acids, and the usual method is 'cracking out' with sulphuric acid. Southgate[22] suggests that the quantity of acid used should exceed that needed to neutralise the alkalinity of the liquor by at least 500 mg/l. On being settled in tanks the grease separates out, either floating to the surface or settling on the bottom with the sludge. The settled liquor can then be discharged into the sewers, but often only after neutralisation with an alkali. From the sludge, grease can be removed by hot pressing or by means of a solvent. Aluminium sulphate can be used instead of acid to remove grease from the waste liquors, but this renders the recovery of grease more difficult.

In some areas in the West Riding of Yorkshire (e.g. Bradford, Huddersfield and Halifax), where the sewage contains a large proportion of wool-scouring waste, steps have to be taken to remove the grease at the sewage works. The grease is 'cracked out' with acid in the primary sedimentation tanks and recovered from the sludge. Steps have to be taken to adjust the pH to within reasonable limits before the settled sewage receives biological treatment.

In some cases the natural grease is not removed from the wool before it is spun and woven into pieces, in other cases oil is added to facilitate these processes. Therefore the woollen pieces have to be

Effects of Trade Wastes

scoured. The waste liquors produced are similar to those produced from the scouring of raw wool but usually somewhat weaker.

Other waste liquors produced in woollen manufacture are from the milling processes, in which the pieces are beaten in a weak alkaline solution.

Waste dye liquors are also produced. Their strength can be reduced by chemical coagulation with alumino-ferric, or copperas and lime, followed by sedimentation before discharge.

LEATHER MANUFACTURE

Tanning is the process of converting the skins of animals into the tough, non-putrescible material known as leather. The processes involved are complex and varied, depending on the type of leather to be produced, and give rise to a variety of waste liquors containing much organic matter. The most satisfactory method of disposing of these waste liquors is by treatment at a sewage works of adequate capacity in admixture with large quantities of domestic sewage.

The main processes are as follows.

(a) Soaking. The skins are soaked in water to remove blood, dirt, dung and curing agents. The waste liquors contain much organic matter and suspended solids.

(b) Liming. The skins are placed in pits or drums containing lime and sodium sulphide or amines to soften the hair and thus facilitate its removal from the skin. The waste liquors are very alkaline and contain large amounts of sulphide, much organic matter and a high proportion of ammoniacal and organic nitrogen. The suspended solids content is high and largely of a fibrous nature.

(c) De-liming, puering, bateing and drenching. After the hair has been removed, the skins which are to be made into heavy leather are de-limed with dilute solutions of acid. Those to be made into light leathers are usually softened by processes known as puering and bateing, which is effected by the action of enzymes. The skins are often subjected to another process known as drenching, which neutralises the alkalinity and cleanses the skins. Most of the waste liquors from these processes are weakly acidic.

(d) Vegetable tanning, effected in pits or drums containing vegetable tanning liquors which are made from barks, woods, nuts and fruits; these have a pyrogallol or catechol base. The

waste liquors are somewhat acidic and contain large amounts of organic matter.
(e) Chrome tanning. The skins are placed in pits or drums containing a solution of a chromium salt, usually basic chromium sulphate. The waste liquors contain chromium which could have an inhibitory effect on the biological processes at the sewage works.
(f) Finishing. The tanned skins are treated with emulsions of sulphonated oil in water, or impregnated with oils and greases. The waste liquors contain much oil and grease.

Some skins are given an oil tannage to produce the chamois type of leather; the waste liquors contain oil and are alkaline.

The use of synthetic tanning liquors made from substances such as formaldehyde, cresols and naphthalene has become more prevalent in recent years.

Leather is sometimes dyed, with the consequent production of waste dye liquors.

It is desirable to give some pre-treatment to waste liquors from leather manufacture. Hair and large pieces of solid matter should be removed by screening to prevent blockages in the sewers. Other suspended solid matter can be eliminated by sedimentation, and judicious mixing of the lime liquors and vegetable tanning liquors will produce some precipitation, thus reducing their strength. If chrome tanning liquors are mixed with the alkaline lime liquors, some of the chromium will be precipitated. Liquors containing large amounts of lime can cause trouble in sewers, as carbonation can take place. This results in a build up of carbonate on the inside of the sewers, which reduces the cross-sectional area and eventually causes a serious blockage.

A process has been developed for the removal of sulphides from tannery effluents[3]. The removal is effected by aerating the liquor in the presence of a manganese salt which acts as a catalyst.

Waste liquors from the finishing processes should be passed through oil traps before discharge to sewers, preferably after treatment with acid or aluminium sulphate.

FELLMONGERING

This entails the removal of wool from sheepskins. The skins are first soaked in water and then treated with lime and sulphide, either by soaking in pits or by application as a paste. After the wool has been removed the skins are soaked in lime water or in a weak acid

Effects of Trade Wastes

solution. The waste liquors produced from fellmongering are similar in character to the lime-yard liquors in the tanning industry. They may contain arsenic, as arsenic sulphide is sometimes used instead of sodium sulphide.

SLAUGHTERHOUSES

Waste liquors from slaughterhouses contain blood, dung, urine, hair, stomach contents and waste from the production of sausage skins, etc., i.e. much organic matter. They have a high 4 h permanganate value and B.O.D. and can place an appreciable load on the sewage works. To prevent blockages it is desirable to screen the liquors before discharging into the sewers. Anaerobic digestion has been used to reduce their B.O.D.

BREWING[22a] AND DISTILLING INDUSTRIES[22b]

These industries produce a variety of waste liquors, the main types being:

(a) Waste liquors from the steeping of grain in the preparation of malt. These have a fairly high permanganate value and can cause formation of growths on percolating filters.

(b) Brewery waste liquors consist mainly of water used for washing casks, bottles, brewing vessels and floors. These contain an appreciable quantity of yeasts and carbohydrates, and often of detergents. They can cause bulking of sludge in an activated sludge plant, and also formation of growths on percolating filters. They are usually acidic, except where an alkaline cleansing agent has been used. Tidswell[23] stated that the volume of waste liquors produced from brewing is between 8 and 12 times the volume of beer produced.

(c) In the distilling industry, where cereals are used the preliminary processes produce waste liquors similar to the brewing industry, but in addition there is the residue from the distillation process. This waste contains much organic matter and can put a heavy load on a sewage works. It is desirable to prevent fluctuations in the discharge of these residues into the sewers, and a balancing tank should be provided. Where sewage containing these liquors is treated on percolating filters, alternating double filtration has often to be carried out to prevent the formation of growths on the filters.

DAIRIES[23a]

The waste liquors discharged from dairies consist mainly of

(a) Wash waters of the milk bottling and distribution departments produced from washing of bottles, cans, plant, utensils and floors;
(b) Waste liquors from butter manufacture which contain butter-milk, whole milk and cream, together with water used for washing of floors and utensils;
(c) Waste liquors from the manufacture of cheese, including the washing waters.
(d) Waste liquors from the manufacture of yoghurt.

All these liquors have a high B.O.D. and contain appreciable amounts of carbohydrates, the washing waters can also contain alkalies or detergents. The volume of waste liquors produced is on average about twice the volume of milk processed.

Where dairy wastes constitute only a small proportion of the sewage to be treated very little trouble is encountered, but a large proportion of dairy waste can cause bulking of sludge in an activated sludge plant or the formation of growths in percolating filters. In the latter case alternating double filtration would have to be used. On those sewage works which have primary sedimentation tanks giving long detention periods, the presence of dairy wastes in the sewage can result in the problem of rising sludge in these tanks, due to the highly putrescible nature of this type of waste.

Dairy wastes should be run into sewers at a uniform rate of flow, which makes the provision of a balancing tank desirable, but care should be taken not to make the balancing tank larger than is absolutely necessary, to avoid the possibility of the waste becoming septic.

The strength of the washing waters can be reduced if steps are taken to ensure that only the minimum of milk runs to waste, e.g. by adequate drainage of milk cans. Pre-treatment with alumino-ferric or copperas and lime, followed by sedimentation, can be adopted to reduce the strength of the waste liquors. The use of filters constructed from plastics media has also produced encouraging results, as this type of filter is not susceptible to blockages by the formation of growths.

FLAX RETTING

In the flax plant the fibres used for making linen lie in bundles over a central wooden core and have to be softened by 'retting' before they can be removed by mechanical means. The usual method adopted is anaerobic retting: the flax is allowed to stand for some hours in warm water tanks; this 'leach' liquor is then run off and replaced by fresh water, and 'retting' takes place. During this period part of the 'retting' liquor is again run off and replaced with fresh water. An anaerobic condition is set up in the tanks, and finally all the 'retting' liquor is run off to waste.

Both the 'leach' and the 'retting' liquors contain much organic matter, and also potash and phosphates. Isaac[24] stated that over 58 000 l of water may be used per tonne of flax retted (13 000 gal/ton); the 'leach' liquors have a B.O.D. of about 1 100 mg/l, the 'retting' liquors of over 2 000 mg/l. Where the proportion of these waste liquors to the volume of sewage to be treated is not large, very little trouble should be experienced at the sewage works. The main difficulty lies in the fact that retting tends to be a seasonal process, so that the load on the sewage works fluctuates over the year.

CHEMICAL MANUFACTURE

The manufacture of chemicals produces a wide variety of waste liquors. Many of them have a high acidity or alkalinity, and they may contain substances such as phenols, nitro-compounds, amino-compounds, aldehydes and heavy metals. Some of them can have a serious inhibitory effect on the biological processes at a sewage works, and pre-treatment is often necessary. The nature of the waste liquors from chemical manufacture is so varied that each case has to be dealt with according to its individual composition.

The range of products manufactured by the chemical industry is continually increasing. Many of them have brought difficult treatment problems in their train. Not the least of these problems are those attendant upon the continual advances in the production of pesticides. A method of dealing with these types of wastes was developed at Fisons Pest Control Ltd., Harston[25, 26]. In dealing with wastes containing weedkillers such as dinitro-o-cresol and substituted phenoxyacetic acids, fungicides such as copper compounds and phenyl mercury salts, and insecticides such as D.D.T. and organic phosphorus compounds, it was found that the wastes had to be neutralised, filtered through sand and treated with activated carbon before they could be treated on percolating filters.

METAL INDUSTRIES

The main waste liquors in these industries come from electroplating and anodising, case hardening, pickling and engineering processes.

(a) *Electro-plating and anodising*—The waste liquors from these processes consist mainly of the water used to wash the articles after they leave the electrolytic and cleaning baths, and are accordingly weak solutions of the electrolytes and cleaning liquors. The strength of these liquors can be considerably reduced by adequate drainage of the articles before washing. In addition to these wash waters the spent contents of the electrolytic and cleaning baths are run off from time to time. The waste liquors from these processes are predominantly acidic and contain copper, nickel, iron, chromates and cyanides in varying amounts, all of which, if present in sufficient quantity, could have an inhibiting effect on the biological treatment processes at the sewage works. Wherever necessary they should be reduced to within reasonable limits by the methods described previously in this chapter. If the cyanides are to be removed by chlorination, the liquors containing nickel salts must be kept separate from the liquors containing cyanides before chlorination. Otherwise nickelocyanides will be formed which will severely retard the action of the chlorine on the cyanides.

(b) *Case Hardening*—In case hardening the metal is dipped in hot molten sodium cyanide and then placed in cold water. This water thus contains cyanides and should only be discharged into the sewers if the cyanide content is within tolerable limits; if it is too high the water will have to be pre-treated.

(c) *Pickling*—The heating of iron or steel causes a layer of scale to form on the surface consisting of oxides of the metal, which has to be removed before further work can be carried out. One of the means of accomplishing this is to pickle the metal in a bath of dilute sulphuric or hydrochloric acid or phosphoric acid, after which it is washed in water. The strength of the acid can vary from 2 per cent to 15 per cent solution. The waste waters fall into two categories:

1. Spent pickle liquors, which are strongly acidic and contain ferrous salts in solution and appreciable amounts of scale sludge.
2. Wash waters, similar in nature to the spent pickle liquors but much weaker.

If these liquors are discharged into sewers and are present in too high a proportion compared with the total volume of sewage, the high acidity can cause corrosion of metals and concrete, and have an adverse effect on the treatment processes at the sewage works. The iron content may cause clogging of percolating filters and diffuser plates in activated sludge plants. In these cases the liquors should be given pre-treatment before discharge: lime treatment followed by sedimentation will neutralise the acidity, precipitate much of the iron and remove the scale sludge.

When copper or copper alloys are heated, a scale consisting of oxides of the metals is formed on the surface, which is often removed by pickling the metal in a bath containing a solution of sulphuric acid, followed by washing the metal with water. The waste liquors produced are

1. Spent pickle liquor, which is strongly acidic and contains sulphates of the metals and scale sludge.
2. Wash waters, i.e. weak solutions of the pickle liquor.

These liquors can cause trouble in the sewers and at the sewage works. The high acidity will cause corrosion of metals and concrete and, together with the metals present, have an inhibitory effect on the biological treatment processes. It is advisable to pre-treat these liquors. Dosage with lime followed by sedimentation will neutralise the acidity, precipitate copper and other metals and remove the suspended solid matter. Electrolysis can be used to treat the stronger liquors.

(d) *Engineering processes*—The main waste from an engineering workshop is oil. This may be either in the form of neat oil or an emulsion of solubilised cutting oil. The source of neat oil is usually leakage from drums or drainage from machinery. This can cause blockages in sewers by causing adhesion of the solids, and will also form oily scum on the surfaces of tanks at the sewage works. As far as possible this oil should be prevented from entering the sewers, and where practicable the drains collecting such drainage and leakage should be connected to a sump from which the oil can be recovered. Where oil falls on floors where it is likely to be washed into drains by rain, an oil trap should be constructed on the drain above the point of connection to the main sewer.

Aqueous emulsions of cutting oil present a different problem, as the oil cannot be separated from the emulsion without chemical treatment; sulphates of iron and aluminium have been found suitable for this purpose. The aqueous layer which separates out after such treatment can be discharged into the sewer.

CANNING INDUSTRY

The canning of vegetables and fruit results in the production of large volumes of waste liquors. These consist mainly of washing waters but also include spillage and drainage from solid wastes such as skins and shells. The liquors have a high organic content and contain much suspended solid matter. Many of the processes in this industry are seasonal, and there is great fluctuation in the volume and nature of the waste liquors. They can be treated efficiently at a sewage works of adequate capacity, provided they do not constitute too great a proportion of the total volume of sewage. They should be passed through fine screens before discharge to prevent blockages in the sewers, and their strength can be reduced by pre-treatment consisting of chemical coagulation and sedimentation. Aluminium sulphate or copperas and lime are considered to be the most suitable coagulants for this purpose. Plastics media filters can also be used to reduce the strength of these liquors.

The practice, introduced in recent years, of marketing some vegetables such as potatoes and carrots in packages, after washing, has resulted in the production of waste liquors which have a high suspended solids and high organic content, which can place an appreciable load on sewage works.

FARM WASTES[26a]

These wastes now come under the heading of 'trade effluent' under the provisions of the Public Health Act, 1961, Part V, and are also subject to the provisions of the Rivers (Prevention of Pollution) Acts, 1951 and 1961, in England and Wales. Although they resemble sewage in composition, in general they are very much stronger. Thus, the drainage from cowsheds may have a B.O.D. as high as 4 330 mg/l; piggery effluents are often much stronger, ranging from 1 275 to 13 260 mg/l; silage liquor, whose production is seasonal, is exceedingly strong, varying in B.O.D. from 12 550 to 66 400 mg/l. Other effluents to be found on farms include waste waters from dairies, drainage from battery-hen houses, and pea-vining liquors; and the present trend towards washing packaged vegetables on the farms is contributing to the effluent problem.

Methods of treatment or disposal include:

(a) Discharge to public sewers. This is a conventional and convenient method if the sewage works is not too far away and is big enough to deal with such strong wastes.
(b) Aerobic biological treatment at the farm. This treatment in percolating filters with recirculation has been used for piggery wastes, but the method is rather expensive and a very high recirculation ratio is needed. Alternatively a high-rate activated sludge process can be used, e.g. an oxidation ditch[27], followed by distribution over land.
(c) Anaerobic digestion[28]. This method is not often used but it does

Effects of Trade Wastes

reduce the strength of the wastes and gives useful gas and fertiliser.
(d) Disposal on land. This is a cheap method and can be used successfully where sufficient suitable land is available e.g. on large dairy farms. There is rarely sufficient land available on those farms which are engaged almost wholly on pig-rearing. Silage liquors can be disposed of on land providing they are first diluted.
(e) Use of soakaway. This is a common method for disposing of silage liquors, but care must be taken that streams or underground water supplies are not polluted.

Other processes from which trade waste liquors are produced include paper coating and staining, tripe dressing, coal washing, plastics and synthetic resin manufacture, sugar beet processing and the manufacture of rubber products, paint, varnish, glue, size and gelatine.

CHARGES FOR TREATING TRADE WASTES AT SEWAGE WORKS

A local authority can charge a manufacturer for the treatment of trade waste accepted into the sewers. In making such charges regard must be made to their volume and composition and to any additional expenditure incurred or likely to be incurred by the sewerage authority in connection with the reception or disposal of the trade waste. This should always be the main criterion in calculating charges for the reception of trade wastes into the sewers. On the one hand the charge must be fair and reasonable to the manufacturer, but on the other hand the manufacturer must not expect the rate-payer to subsidise the cost of disposing of his wastes. As there are such wide variations in conditions and costs of sewage treatment throughout the country it has not been found possible so far to formulate a standard scale of charges.

A rough yardstick for assessing the effects of trade wastes on a sewage works is given by expressing the trade wastes in terms of equivalent populations.

The five-day *per capita* B.O.D. in Britain is now considered to be 0·073 kg (0·16 lb) per day, so the population equivalent is now given by:

$$\text{Population equivalent} = \frac{\text{five-day B.O.D. (mg/l)} \times \text{flow (m}^3\text{)}}{0\cdot073 \times 10^3}$$

This population equivalent can be of assistance in calculating the charges which should be imposed for the reception of trade waste into the sewer. Where necessary the *per capita* B.O.D. of the existing flow of sewage can be worked out for any individual sewage works.

At the Greater London Council's Mogden Sewage Works[29] the following formula, known as the 'Mogden Formula', has been used for fixing the charge for the admission of some trade wastes into the sewers:

$$\text{Charge (pence*/1 000 gal)} = 1 + \frac{M}{75} + \frac{S}{60}$$

where M = McGowan strength of the settled trade waste (parts per 100 000)

S = Suspended solids (parts per 100 000)

60 = Factor based on cost of sludge disposal

75 = Factor based on cost of biological treatment

Griffiths and Kirkbright[30] give the following generalised form of the Mogden Formula:

$$C = V + \frac{M_t}{M_s} B + \frac{S_t}{S_s} S$$

where C = charge (pence* per 1 000 gal of trade effluent)

V = cost (per 1 000 gal of D.W.F. of sewage treated) of preliminary treatment and miscellaneous operations.

B = cost (per 1 000 gal of D.W.F. of sewage treated) of aerobic biological purification.

S = cost (per 1 000 gal of D.W.F. of sewage treated) of sludge disposal.

M_t = McGowan strength of settled trade waste
M_s = McGowan strength of settled sewage
S_t = Suspended solids content of trade effluent
S_s = Suspended solids content of sewage

} All expressed in same units

It will be seen that the cost of trade waste treatment is made up of three elements: preliminary treatment, biological treatment, and sludge disposal. Such costs will vary from place to place.

The Mogden Formula is not regarded as suitable for the type of

* Old pence.

Effects of Trade Wastes

industrial sewage and trade waste dealt with in Birmingham. There the charges are based upon the costs of sedimentation, biological oxidation and sludge treatment, with additional charges for the cost of pumping[31]. There are special maximum charges for galvanising and ferrous metals pickling wastes (2·78 pence* per 1 000 gal), for wastes from metal finishing, electroplating and non-ferrous metal pickling (8·34 pence* per 1 000 gal), for gas liquor, and for other unusual wastes.

Each manufacturer is charged in accordance with the degree to which his type of waste involves sedimentation, biological oxidation and sludge disposal methods.

In the year ending 31st March 1962, the costs of these three treatment processes at Birmingham were as follows:

	Cost (pence* per 1 000 gal)
Sedimentation	1·39
Biological treatment	2·78
Sludge treatment	1·39
Total cost	5·56 (2·32 new pence)

Food industry wastes similar to sewage are costed at the same maximum rate as sewage, i.e. 5·56 pence* per 1 000 gal. The permissible chemical limits for these discharges are assumed to be the same as the average for Birmingham sewage, i.e. B.O.D. 322 mg/l, 4 h permanganate value 125 mg/l and suspended solids 400 mg/l. If by pretreatment a trader succeeds in reducing the pollution load of his effluent below these limits, he is given credit for it in a proportionately lower charge. Thus, if a cannery waste had a suspended solids content of 200 mg/l instead of the prescribed 400 mg/l, the portion of the charge allotted to sludge treatment would be halved. There is an additional charge of 1·6 pence* per 1 000 gal for the use of the Corporation's sewers. Charges are reviewed annually and varied accordingly.

In a paper given to the Institute of Sewage Purification Conference Kershaw[32] explained a modified form of the Mogden Formula used at Slough, as follows:

$$\text{Charge (pence*/1 000 gal)} = A + \frac{M}{B} + \frac{S}{C}$$

where M = McGowan strength of the trade waste

S = Suspended solids content of the trade waste

A = Cost per 1 000 gal for preliminary treatment of the raw sewage, including general expenditure

* Old pence.

B = Factor based on cost per 1 000 gal for biological treatment of the tank effluent

C = Factor based on cost of treating the solids content of the raw sewage.

It will be seen that the charges can be varied in accordance with changes in the cost of sewage treatment.

It should be remembered, however, that some trade wastes contain substances which interfere with the normal processes at the sewage works and create difficulties disproportionate to their oxygen demand and suspended solids content. Consideration has often to be given to this point when assessing the charges. Some local authorities take the view that, as the municipal sewers are being used to transport the trade wastes to the sewage works, this should be taken into account when calculating the charges.

If a sewage works is to be operated efficiently, frequent inspection of discharges from manufacturers' premises should be made by the local authority concerned, to ensure that the conditions regulating the discharges are being observed.

REFERENCES

1. *Public Health (Drainage of Trade Premises) Act,* 1937. London; H.M.S.O.
2. *Public Health Act,* 1961. London; H.M.S.O.
3. HUMPHREYS, F. E. and BAILEY, D. A., Tannery effluents: what are the problems, *Wat. Pollut. Control* 68 (1969) 93
4. *Water Pollution Research,* 1952. London; H.M.S.O. 1953
5. PETTET, A. E. J. and MILLS, E. V., Biological treatment of cyanides, with and without sewage, *J. appl. Chem. Lond.* 4 (1954) 434–44
6. *Water Pollution Research,* 1953. London; H.M.S.O. 1954
7. WALDMEYER, T., Treatment of formaldehyde wastes by activated sludge methods, *J. Proc. Inst. Sew. Purif.* 1 (1952) 52; see also SINGLETON, K. G., *J. Proc. Inst. Sew. Purif.* 6 (1965) 498–506
8. BURGESS, S. G., Analysis of trade waste waters, in *Treatment of trade waste waters and prevention of river pollution,* Isaac, P.C.G. (Ed.), 1957. London; Contract. Rec.
9. RUGAL, H. T., Copper in sludge digestion: effect of copper on the operation of the Kenosha plant, *Wat. Sewage Wks* 93 (1946) 316
10. LUMB, C., Effects of synthetic detergents on sewage purification: a summary of current knowledge, I. *Wat. Sanit. Engr* 3 (1952) 7, 25
11. Symposium on New problems in sewage treatment, with special reference to synthetic detergents. *J. Proc. Inst. Sew. Purif.* 1 (1948) 100
12. HURLEY, J., Influence of synthetic detergents on sewage treatment, *J. Proc. Inst. Sew. Purif.* 3 (1950) 249
13. HURLEY, J., Experimental work on the effect of synthetic detergents on sewage treatment, *J. Proc. Inst. Sew. Purif.* 4 (1952) 306
14. LUMB, C., Experiments on the effects of certain synthetic detergents on the biological oxidation of sewage, *J. Proc. Inst. Sew. Purif.* 4 (1953) 269; *Wat..Sanit. Engr* 4 (1954) 305
15. *Control of radioactive wastes,* Cmd. 884, 1958. London; H.M.S.O.
16. *Radioactive Substances Act,* 1960. London; H.M.S.O.
17. *Recommendations of the International Commission on Radiological Protection.* I.C.R.P. Publication 6 (as amended 1959 and revised 1962), 1964. Oxford; Pergamon Press

17a. LITTLE, A. H., The treatment and control of bleaching and dyeing wastes, *Wat. Pollut. Control* 68 (1969) 178
18. *Water Pollution Research*, 1952, 1953, 1954. London; H.M.S.O.
19. *Liquor Effluents Research Committee*, 1st Report, 1927. London; Institution of Gas Engineers
20. Institute of Sewage Purification, Memorandum on the disposal of gas works effluents, *J. Proc. Inst. Sew. Purif.* 2 (1949) 104; see also *J. Proc. Inst. Sew. Purif.* 1 (1956) 5
20a. WALDMEYER, I., Use of synthetic plastic media for the biological treatment of paper mill wastes, *Paper Maker & Brit. Pap. Trade J.*, Nov., 1964
21. HUGHES, J. W., Industrial waste treatment at a viscose rayon factory, *J. Proc. Inst. Sew. Purif.* 4 (1951) 406
22. SOUTHGATE, B. A., *Treatment and disposal of industrial waste waters*, 1948. London; H.M.S.O.
22a. AULT, R. G., An approach to the problem of brewery effluents. *Chemy Ind.* 4 (1969) 87
22b. ASKEW, M. W., Plastics in waste treatment, *Process Bio-chem.*, 1 (1966) 483–86, 492 and 2 (1967) 31–34
23. TIDSWELL, M. A., Treating sewage containing brewery waste, *Contract. Rec.* (2nd March, 1960) 15, 30
23a. FISHER, W. J., Treatment and Disposal of Dairy Waste waters: a review, *Dairy Sci. Abstr.* 30 (1968) 567
24. ISAAC, P. C. G., *Public Health Engineering*, 1953. London; Spon
25. SHARP, D. H. and LAMBDEN, A. E., Treatment of strongly bactericidal trade effluent by activated charcoal and biological means, *Chemy Ind. Rev.* No. 39 (1955) 1207–16
26. LAMBDEN, A. E., The Fisons Pest Control Ltd., trade effluent treatment plant, *J. Proc. Inst. Sew. Purif.* 2 (1959) 174–81
26a. GIBBONS, J., Farm waste disposal in relation to cattle, *Wat. Pollut. Control* 67 (1968) 622
27. SCHELTINGA, H. M. J., Aerobic purification of farm waste, *J. Proc. Inst. Sew. Purif.* 6 (1966) 585
28. ANON., Farm effluent plant produces gas for domestic heat and power, *Wat. Waste Treat. J.* 9 (1963) 434
29. TOWNEND, C. B. and LOCKETT, W. T., *J. Proc. Inst. Sew. Purif.* 1 (1947) 48
30. GRIFFITHS, J. and KIRKBRIGHT, A. H., Charges for treatment of trade wastes (Middlesex). *J. Proc. Inst. Sew. Purif.* Pt. 4 (1959) 505–21
31. ASHER, S. J. A., Trade effluent control in Birmingham, *J. Proc. Inst. Sew. Purif.* 6 (1962) 497–520
32. KERSHAW, M. A., Some trade effluent problems in a rapidly developing town, *J. Proc. Inst. Sew. Purif.* 3 (1955) 242

Chapter 11

SMALL SEWAGE TREATMENT PLANTS

The most satisfactory method of dealing with domestic sewage is by discharging it to the main sewerage system for treatment at municipal sewage works, but in the case of some isolated houses or groups of houses, especially in rural areas, this is not a practical or economic proposition due to the length of 'dead' sewers which would be involved. In these cases steps have to be taken to deal with the sewage in the vicinity of such small communities. The building of new houses in unsewered areas should, however, be discouraged, unless the houses have to be built in that place for a specific purpose. The reason being, that however well designed the treatment plant may be, there is a tendency, if it is under private control, for it to be neglected, with a consequent deterioration in the quality of the effluent. As in many cases these small plants discharge into watercourses which are no more than ditches, a multiplicity of such plants in an area can result in very unsatisfactory conditions arising in the local watercourses.

The Ministry of Housing and Local Government has issued two publications[1, 1a] on the subject. There is another publication on the same subject issued by the Council for Codes of Practice–British Standards Institution in 1956[2] in which recommendations are made regarding capacities and design of plants suitable for dealing with the sewage from isolated communities.

In the past the prevalent method was collection in a cess-pool. A cess-pool is a watertight covered tank, ventilated to allow the gases to escape, in which the sewage is stored so that it can eventually be collected by a tank waggon and conveyed to the sewage works for disposal. It should be remembered that cess-pool liquors tend to be very strong; often three times as strong as an average sewage, and can be septic in character. If these liquors have to be treated at a small sewage works, they can be a source of trouble, and chlorination may be necessary. This can be done by dosing with hypochlorite, but care should be taken that an excess is not used which would interfere with the aerobic biological processes. One of the faults with cess-pools is that they can, in some cases, be anything but watertight and can be a source of pollution to nearby watercourses or to underground water.

The means commonly employed today consist of sedimentation followed by biological oxidation, after which the effluent is discharged into a watercourse or allowed to soak away into the land. These plants can be regarded as scaled-down versions of local authorities' large sewage works, but as they obviously will not receive the same attention, some design modifications have to be made so that they will operate efficiently with the minimum of attention. It is essential that surface water be excluded from these small plants, because owing to the fact that the septic or sedimentation tanks are comparatively small, the great increase in flow in times of storm would cause excessive turbulence, with the result that appreciable quantities of solid matter would be carried forward to the secondary treatment section of the plant and thus cause clogging of filters, etc. Screens and grit tanks are often dispensed with because they require too much attention. If there is insufficient fall through the site and pumping has to be resorted to, this should take place after sedimentation to prevent clogging of the pumps. It should always be remembered that in rural areas there is a tendency for less water to be used than in towns, especially where premises are served by a private water supply. The sewage, therefore, may be very strong, e.g. B.O.D. of 600–750 mg/l instead of about 300–500 mg/l as in urban areas.

SEDIMENTATION

Sedimentation is usually carried out in septic tanks, i.e. covered sedimentation tanks having a large capacity in relation to the daily flow of sewage. They thus provide sludge storage capacity and cause some reduction in the volume of sludge to be disposed of, owing to the anaerobic digestion that takes place during the fairly long period the sludge remains in the tank. This is necessary because it is not usually practicable to desludge the tanks at frequent intervals, and land suitable for sludge disposal is not often available. The best means of desludging these tanks is by means of a tank waggon (a gully emptier is often used for this purpose), the sludge being taken away for disposal on a suitable site. The extra capacity provided in a septic tank enables it to absorb much better the peak flows of sewage which might cause turbulence in a small tank and leave excessive suspended solids in the tank effluent. It is suggested in the Ministry's Memorandum that septic tanks rather than ordinary sedimentation tanks should be provided for sewage plants serving a population of less than 100 persons. Many local authorities today insist that the minimum capacity of a septic tank shall be 2 730 l (600 gal), even for a plant serving one house.

SECONDARY TREATMENT

The secondary treatment can consist of percolating filters, land treatment or activated sludge plant.

PERCOLATING FILTERS

For satisfactory results to be obtained from these small plants an efficient system of distributing the tank effluent on to the filter is essential, and difficulties are encountered owing to the fact that at certain times during the day the flow of sewage entering the plant will fall to a mere trickle. The fixed type of distributor, consisting of a system of channels fitted over the filter media, is unsatisfactory unless some kind of dosing mechanism such as a tipping tray is incorporated. For the larger plants a circular filter with a rotating distributor is desirable, but this should be fitted with a dosing siphon, tipping trough or waterwheel to ensure satisfactory rotation of the distributor. For satisfactory performance the depth of media should never be less than 1·2 m (4 ft), and there should be adequate ventilation. It is sometimes desirable to cover these filters to prevent leaves and rubbish blowing on to them; wire netting stretched on frames, concrete slabs or wooden sleepers can be used for this purpose but spaces must be left between the slabs or sleepers to allow free access of air to the filter.

Provisions must be made for removing the cover, to allow maintenance of the distributor to be carried out. In view of the strong character of septic tank effluents, the dosage on the filter should not exceed 270 l/m^3 day (45 gal/yd^3 day), and a lower dosage may be found to be necessary in some cases.

The desirability of providing a humus tank is a debatable point, as it requires frequent desludging and any neglect of this will result in serious deterioration in the nature of the effluent. It is not usual to provide humus tanks for the smaller plants. Some of the humus can be removed by running the filter effluent over grass plots, and sand filters have been used successfully in some cases.

LAND TREATMENT

Where a sufficient area of suitable land is available the secondary treatment can be carried out by surface or sub-surface irrigation. The land needs to be of good porosity; a light loamy soil is the best, peat and clay are virtually useless for this purpose.

SURFACE IRRIGATION

The tank effluent is distributed on to the land from channels constructed for the purpose, the distribution being controlled by simple hand-operated penstocks. If the land is sufficiently porous 17–25 m^2 (20–30 yd^2) of land per house should be adequate. The laying of field drains at a depth of approx. 0·9 m (3 ft) below the surface and discharging into a watercourse, and the digging of grips on the surface into which the sewage can be run, will be beneficial in cases where porosity is not very good; this is, of course, land filtration rather than irrigation.

At least half the area of land must be 'resting' while the remainder is being used for sewage treatment.

SUB-SURFACE IRRIGATION

The tank effluent is run into a system of open-jointed field tiles laid in trenches about 0·6 m (2 ft) deep and surrounded by clinker or other filter media. This is really a soakaway system, but some purification is effected if the field tiles are not laid deep enough to prevent the access of air to the sewage. Where the land is not porous two lines of field tiles can be laid in the same trenches. The sewage is run into the top line which is about 0·6 m (2 ft) above the bottom line, the space between being filled with coarse sand. The bottom line of field tiles discharges into a watercourse. This system is not entirely satisfactory and is expensive to construct.

The Ministry's Memorandum states that land used for the treatment of sewage in any of the above ways should not be within 90 m (300 ft) of any water supply source, and should be geologically below it.

ACTIVATED SLUDGE PLANTS

In the past, activated sludge plants have rarely been used for a small community, since they require frequent and skilled attention and the disposal of surplus activated sludge presented a difficult problem, but in recent years certain manufacturers have developed small plants which require less attention.

The package type of extended aeration plant (see p. 97) has, recently been used for the treatment of small volumes of sewage. Tertiary treatment, preferably on grass plots, should be given to the effluents from such plants and it is important to exclude surface water from the plants. Surplus activated sludge has to be removed from time to time and the plants should not be left for long periods without inspection and attention.

Another type of plant which has come into use in recent years for the treatment of sewage from small communities, and indeed for large villages, is the 'Pasveer Oxidation Ditch'[3]. This consists of an oval shaped ditch, the sides of which are often protected by special sheeting or by concrete slabs. A rotor similar to the one used in Kessener aeration plants is used to provide aeration and to maintain a flow velocity in the ditch of about 0·3 m/s (1 ft/s). The capacity is usually from 1–2 days D.W.F. The sewage is normally introduced without prior sedimentation. Final settlement of the mixed liquor can be effected in an external sedimentation tank, or in a secondary ditch, or by providing three lanes in the ditch, two of which are used alternately, by means of automatically controlled penstocks, as aeration or settlement lanes. These plants can be constructed fairly quickly and construction costs are low.

'The Pasveer Oxidation Ditch' like the package extended aeration plants can be used as a fairly quick means of providing additional biological oxidation capacity for overloaded sewage works.

MINISTRY PUBLICATION ON ACTIVATED SLUDGE SEWAGE-TREATMENT INSTALLATIONS[3a]

The memorandum states that in designing these small long-period activated sludge plants consideration should be given to the range of activated sludge solids which should be present under operating conditions. This is expressed as the weight of B.O.D. which will be applied per day in relation to the unit weight of activated sludge solids present and is referred to as B.O.D./sludge loading factor. The Report suggests that this should be within the range 0·05–0·15 B.O.D. per day per gramme of activated sludge solids; this generally corresponds to activated sludge solids concentrations of 2 000 to 6 000 mg/l. In designing final settlement units, it is suggested that the surface overflow rate at maximum flow should not exceed 22 000 l/m^2 day (450 gal/ft^2 day) and that the maximum solids loading per unit area of surface should not exceed 120 kg/m^2 day (25 lb/ft^2 day).

The memorandum also makes the following recommendations:

Extended aeration plants
B.O.D. loading should not exceed 240 mg/l (15 lb/1 000 ft^3) of aeration capacity per day.

Contact stabilisation plants
The B.O.D. loading of the first two aeration stages (*i.e.* 'contact' and 're-aeration') should not exceed 480 mg/l (30 lb/1 000 ft^3) of their combined capacity per day.

Oxidation ditches
B.O.D. loading should not exceed 210 mg/l (13 lb/1 000 ft^3) of ditch capacity per day.

Polishing of plant effluent
  (a) *Grass plots.* About 3 m^2 (4 yd^2) of grassland per person per day is normally required.
  (b) *Upward-flow gravel bed clarifiers.* Gravel bed should be 150 mm (6 in) in depth of 5–6 mm ($\frac{3}{16}$–$\frac{1}{4}$ in) gravel. Flow at maximum rate should not exceed 980 l/m^2 h (20 gal/ft^2 h).
  (c) *Slow Sand Filters.* Flow should not exceed 150 l/m^2 h (3 gal/ft^2 h).

OXIDATION PONDS

In the U.S.A., New Zealand and Australia, oxidation ponds have been used to treat sewage from small communities but this system has so far not found favour in the British Isles.

DISPOSAL OF EFFLUENT

After purification the effluent is usually disposed of by soaking it away into the land or by discharging it into a watercourse.

To use a soakaway system the land must be of good porosity; it is usually carried out by running the effluent into a system of open-jointed field tiles laid in trenches and surrounded by clinker or rubble, or into a pit filled with rubble. Wherever practicable, the effluent from the smaller plants should always be disposed of in this manner, providing there is no danger of any damage being caused to underground water sources.

If the effluent is discharged into a watercourse it will usually be found that the River Authority concerned will require that it should be of at least the standard recommended by the Royal Commission for sewage effluents (i.e. B.O.D. not to exceed 20 mg/l and suspended solids not to exceed 30 mg/l). In addition, suitable provisions will have to be made to enable the River Authority to take samples of the effluent.

Some valuable investigational work on the treatment of sewage from small communities was carried out by Truesdale et al.[4] of the Water Pollution Research Laboratory.

REFERENCES

1. *Ministry of Housing and Local Government, Memorandum on principles of design for small domestic Sewage Treatment Works*, 1953. London; H.M.S.O.
1a. *Ministry of Housing and Local Government. Operation and management of small sewage works* 1965. London; H.M.S.O.
2. *Small Domestic Sewage Treatment Works,* General Series. The Council for Codes of Practice, British Standards Institution, British Standard Code of Practice C.P. 302.100 (1956)
3. *The oxidation ditch.* Research Institute for Public Health Engineering, T.N.O. The Hague, Holland. Obtainable from Whitehead & Poole Ltd., Radcliffe, Manchester
3a. *Ministry of Housing and Local Government. Technical Memorandum on Activated Sludge Sewage-treatment installations providing for a long period of aeration,* 1969. London; H.M.S.O.
4. TRUESDALE, G. A., BIRKBECK, A. E. and DOWNING, A. L., The treatment of sewage from small communities, *J. Proc. Inst. Sew. Purif.* 1 (1966) 34

Chapter 12

TRENDS IN THE FIELD OF WATER POLLUTION CONTROL

INTRODUCTION

During the last 70 years great progress has been made in the science of sewage treatment, and many changes have been brought about in the design and operation of sewage works. Happily, this progress is continuing, thanks mainly to the inspiration of the Institute of Water Pollution Control and the activity of its members, and to the valuable research work carried out by the Water Pollution Research Laboratory of the Ministry of Technology. In this chapter are outlined some of the paths along which progress is being made, and the lines of approach to find solutions to prevailing problems.

GENERAL

REGIONALISATION

There had been a trend during the past 35 years towards the regionalisation of sewage treatment, that is to say instead of each local authority treating the sewage produced within its own boundaries, an authority is established to treat sewage from a wider area determined by geographical features and not necessarily by municipal boundaries. A number of authorities have already been set up for this purpose, e.g. the former West Middlesex Drainage Authority, whose works are now under the control of the Greater London Council, the Bolton and District Joint Sewerage Board, and the Upper Tame Main Drainage Authority, and there is a prevailing school of thought which favours the setting up of more of these regional authorities. In fact, Section 9 of the Rivers (Prevention of Pollution) Act 1951 gives the Minister power to direct local authorities to act jointly on sewerage and sewage purification by setting up a Joint Sewerage Board. Such powers have, however, rarely been used. Treatment of sewage on a regional basis has the great advantage of qualified technical supervision not always available to a small local authority, and often allows for more trade waste to be treated at the sewage works. Better facilities for long-term research are also often possible. In the existing regional schemes the tendency has been to construct a very large plant to deal with all the sewage from the whole of the drainage area, but this can have the disadvantage that, even though the works produces a good effluent, a great pollution load is placed on the river at one point owing to the large volume of effluent discharged. Moreover, any power failure or major breakdown at the works would result in large volumes of crude or partially treated sewage being discharged into the river. The alternative of having a regional authority controlling a number of medium-sized works is a line of approach which warrants serious consideration for the future, in the light of local circumstances. This course of action would utilise the self-purifying powers of rivers to a greater extent than can one large plant, and would also reduce the risk of pollution due to breakdowns. This is of particular importance in view of the present emphasis on the need to use the water resources of the country as efficiently as possible, including the use of river water for domestic purposes to a greater extent than in the past.

USES OF SEWAGE EFFLUENTS AND RECLAMATION OF WATER FROM THEM[1,2]

Final sewage effluents from aerobic biological plants in Great Britain are generally discharged to watercourses. But in many industrial areas, and also in the drier parts of England (e.g. the south-east) reasonably pure water for domestic and other uses is becoming increasingly difficult to find. Large scale schemes to augment the water supplies in northern and eastern parts of the country (e.g. the Wash, Morecambe Bay and Solway Firth schemes) will take a long time to complete even if Government permission for their implementation is obtained. The reclamation of water from sewage effluents, i.e. the re-use of such effluents, if necessary after further treatment, is, therefore, becoming of importance and much greater utilisation of sewage effluents for industrial and other purposes is probable in the future.

To some extent, the use of sewage effluents is already being practised e.g. as cooling water at many electricity generating stations, as at Coleshill (Birmingham), Croydon, Stoke-on-Trent, St. Helens and Oldham.

In South Africa, a hot dry country, with limited water supplies, sewage effluents are either sand-filtered, or treated in oxidation ponds. At Baltimore, U.S.A., some 545×10^3 m^3/day (120×10^6 gal/day) of sewage effluent were used in 1965 as cooling water at the Bethlehem Steel Company's Works. Sewage effluents, sometimes without any further treatment, are used in various parts of the world (e.g. Israel) for irrigation since they contain a proportion of the fertilising constituents of sewage. About 50 per cent of the N and P originally present will still be present in the effluent as well as most of the potash.

In general some pre-treatment of the activated sludge plant effluents or percolating filter effluents is necessary and this can be done by tertiary treatment or by the application of well-known water treatment techniques, such as micro-straining, chlorination to prevent algal and bacterial growths, and for sterilisation, and chemical flocculation.

The town of Warrington has for many years been using water from the River Mersey (which consists largely of sewage effluents and trade effluents in dry weather) for the various industrial undertakings in that town. Treatment is by screening, coagulation with alum and activated silica, sand filtration and chlorination. The treated water sells at a price much below that of the town's domestic water supply.

The Water Pollution Research Laboratory[2] has described a pilot water-reclamation plant suitable for producing water of reasonable purity from Stevenage sewage effluent. The various stages of treatment (or unit operations) are in the following order:

1. The Banks upward-flow pebble bed clarifier.
2. Use of air in a foaming column to produce foam and so reduce the synthetic detergent concentration of the effluent.
3. Coagulation with ferric salts or aluminium sulphate. If necessary, chlorination is practiced to prevent rising of sludge.
4. Passage through a mechanical flocculator.
5. Use of sludge blanket clarifier.
6. Rapid gravity sand filtration.
7. Final chlorination (this may or may not be necessary).

The sewage effluent is treated at the rate of 45 l/h (10 gal/h). A summary of the results of these processes is given in *Table 17*.

Another method being investigated by the Water Pollution Research Laboratory does indeed involve the use of ozone[3], which it is hoped will become cheaper in the future. The sewage effluent, after micro-straining to remove suspended solids, is treated with ozone and then filtered through sand. The ozonation converts colloidal matter into a form readily removed by sand filtration and also decolourises the effluent and sterilises it. A pilot plant treating up to 10 m³ (2 200 gal) per hour started operation in November 1966. The process is stated to remove completely not only coliform bacteria but also faecal streptococci and viruses.

Table 17. AVERAGE PERFORMANCE OF PILOT PLANT TREATING HUMUS TANK EFFLUENT FROM STEVENAGE SEWAGE WORKS

Constituent	Stevenage humus tank effluent	Banks Clarifier effluent	Foaming column effluent	Final effluent (reclaimed water)	Composition of Stevenage tap water (for comparison)
Total Solids (mg/l)	728	722	700	724	390
Suspended Solids (mg/l)	15	9	6	1	nil
B.O.D. (5-day) (mg/l)	9	6	3	nil	nil
C.O.D. (dichromate value) (mg/l)	63	57	46	33	3
Phosphate (as P) (mg/l)	9·6	9·6	9·5	4·0	0·03
Nitrate (as N) (mg/l)	38	38	38	38	5·0
Chloride (as Cl) (mg/l)	69	–	–	109	20
Detergents (as Manoxol O.T.) (mg/l)	2·5	2·6	0·7	0·6	nil
Ammonia (as N) (mg/l)	4·1	3·8	3·9	2·0	nil
Total hardness (as CaCO) (mg/l)	249	–	–	263	–
Colour (Hazen units)	50	50	50	17	–
pH value	7·6	–	–	7·4	–
Coliform Bacteria (number per ml)	–	–	1343	0·01*	nil

* i.e. only 1 per 100 ml. A satisfactory drinking water should contain only 1–3 coliform organisms per 100 ml.

It will be seen from *Table 17* that the reclaimed water contains a much higher total solids concentration and much more organic matter than tap water. Removal of these would require special water techniques (e.g. activated carbon or evaporation). Although the coliform count is negligibly small, there is still a possibility that harmful enteric viruses might be present and therefore it would be risky to use the reclaimed water for drinking purposes. Moreover, the high nitrate content would make the water unsuitable for small babies as it could give them the disease known as methaemoglobinaemia, which is often fatal. It is desirable that the nitrate content of water used for drinking purposes shall not exceed 0·5 mg/l. The treated water is also high in chloride, derived from the sewage, the water supply, the coagulant used, and also from any chlorine used for sterilisation. Ozone might be used in place of chlorine for sterilisation, but its use in this country is still limited owing to its high cost.

There seems little doubt that the sewage works manager of the future may have, in many instances, to provide a suitable water for industrial and other purposes and so may have to operate a water reclamation plant. This will add a new dimension to his present responsibilities, for he will be supplying a valuable commodity, i.e. water, on a commercial basis.

Eden *et al*[2] give the approximate costs of water reclamation processes as follows; they are largely based on U.S. development work:

Table 18. APPROXIMATE COSTS OF WATER RECLAMATION PROCESSES

Stage	Cost (pence† per 1 000 Imp gal.)
*Sewage Treatment–Primary and Secondary	12
Tertiary	1–3
Foaming	1–2
Coagulation and sand filtration	8
Evaporation	70
Chlorination	1

* U.K. data
† Old pence

VARIATIONS IN THE LOAD ON A SEWAGE WORKS

At most sewage works there are great variations throughout the day, even in dry weather, in volume and strength of the sewage received, and these are usually accentuated by appreciable quantities of trade waste. Such variations can cause difficulties in the operation of a sewage works, especially in the biological oxidation plants, and the position is aggravated by the fact that quite often peak flows coincide with peak strengths of sewage. Stanbridge[4] made a thorough examination of the problem and pointed out that, with domestic sewage alone, in dry weather there is a variation of about 3:1 between high and low flows, and variations in the total load (i.e. the product of volumetric and impurity load) can be of the order of 8:1 to 10:1. His suggestions for equalising the load include provision of primary sedimentation tanks designed to act likewise as balancing tanks, that of separate balancing tanks, dry-weather use of storm water tanks as balancing tanks and provision of automatic distribution chambers on percolating filter installations to add or shut off filter units as the flow rises or falls. He points out that where variations are accentuated by periodical discharges of trade waste, this problem should be solved by controlling this flow at its source. He contends that in activated sludge plants equalisation takes place to a much greater extent than in percolating filters.

Extreme variations in load occur in towns such as holiday resorts and in those rural areas where large caravan sites are situated. In these cases the drainage population is much greater in summer than in winter and can also be much greater on Saturdays and Sundays than during the rest of the week. This results in great seasonal variations in addition to the usual diurnal ones. Seasonal variations can also occur on a small rural sewage works having a large motorway restaurant in its drainage area. This trend could increase and great attention will have to be paid to the design of a sewage works to treat the sewage from such an area.

MECHANISATION AND AUTOMATION AT SEWAGE WORKS

Automatic control of many of the continuous processes and operations at sewage works is already well advanced, especially at some of the newer plants. The control room with its panel of indicating, recording and controlling devices is now an accepted part of the order of things in this age of automation. In the sludge disposal field, especially, many new developments have taken place in respect of the mechanical handling of sludge and in new dewatering techniques. Labour, expensive and difficult to obtain, can be kept to a minimum by the adoption of mechanisation and automation—but, of course, any breakdown in automatic operation can lead to difficulties. To meet these emergencies it is advisable to ensure that plant units can be, as far as possible, kept in operation by manual control until repairs are effected.

There is little doubt that during the next few years even further progress will be made towards more complete automation at the larger sewage works. The provision of highly sophisticated plant at small works attended by semi-skilled or unskilled attendants can, however, at times be a liability, and the installation of this type of plant at these works should be given very serious and careful consideration in the light of local circumstances, before being undertaken.

IMPROVEMENTS TO THE APPEARANCE OF SEWAGE WORKS

It has long been accepted practice in the U.S.A. to make sewage plants attractive to the eye, not only by good design but also by planting shrubs and trees and laying out ornamental lawns and flower beds. The advantages of doing this are being realised in this country; not the least is the incentive it provides for the plant personnel to keep the works clean and tidy. Caution should, however, be exercised regarding any decision to plant trees. These, if situated near the plant units can be a source of trouble. Falling leaves every autumn can make roads and paths slippery and dangerous, the dead leaves can cause blockages on the plant and are a nuisance on the surface of filters. The removal of these dead leaves can be costly and time-consuming. In addition tree roots can cause damage to underground pipe-work and foundations. A possible solution is to limit the trees, preferably evergreens, to the perimeter of the works, so that they act as a screen, and to use small evergreen shrubs on the works itself for ornamental purposes. Large areas of grass can be expensive to maintain in good condition, so that it is imperative when laying out grassed areas to do this in such a manner as will ensure that grass cutting can be carried out as easily as possible. The use of sheep to keep grass short is always worthy of consideration.

Many of the newer works are no longer tucked away in remote spots where they are out of the public eye but are often situated near large centres of population, and it is a curious psychological fact that there are likely to be fewer complaints about smells, etc. from a sewage works if it is effectively designed and laid out. Regarding smells, the growing scarcity of land in England and the need to build some sewage works close to housing estates has made the problem of bad odours an important one. Chlorination is one way of preventing sewage from becoming septic and smelling of hydrogen sulphide. At the Mogden Sewage Works, close to the town of Isleworth, Middlesex, over 70 Air Wick spray-guns are used to control odours[5]. In the U.S.A. ozone had been used successfully to eliminate odours[6]. Sewer cleaning is an important means of locating trouble areas and so eliminating hydrogen sulphide odours. Pre-aeration tanks are sometimes used to oxidise and remove any free sulphide present and so prevent bad smells.

NEED FOR EFFLUENTS BETTER THAN ROYAL COMMISSION STANDARD[7, 8, 9]

Methods for improving final effluents from activated sludge plants and percolating filters so as to achieve standards better than those of the Royal Commission have already been discussed in Chapter 7, and the present-day emphasis on conserving water resources has been referred to earlier in this chapter. It must be remembered that the Royal Commission standards were put forward over 50 years ago when technical knowledge was not so advanced, and for those days they were no doubt adequate. Times have changed since then, however, and something better is needed in many instances. Just as our educational system, our roads and transport, our lighting, our domestic appliances and indeed our whole industrial economy are undergoing, or have undergone, revolutionary changes to meet modern conditions, so our sewage effluents should be tailored to satisfy the challenge of the age and the desire for cleaner rivers. That there is a definite trend in this direction is illustrated by the fact that a number of local authorities are either already producing, or are aiming to produce, final sewage effluents considerably better than Royal Commission. Luton was the first authority to achieve such a distinction (cf. Chapter 7, p.111); by use of micro-strainers or sand filters, a final effluent was obtained similar in quality to a doubtful river water (B.O.D. and suspended solids about 6 mg/l). Since then, several other authorities have agreed to conform to stricter standards, as shown in *Table 19*. Owing to difficulties in maintaining a strict standard the whole of the year, Luton now works to a more lenient standard during the winter.

Table 19. LOCAL AUTHORITIES WORKING TO SEWAGE EFFLUENT STANDARDS STRICTER THAN THOSE OF THE ROYAL COMMISSION

Authority	B.O.D. (5-day) not more than (mg/l)	Suspended solids not more than (mg/l)
Crawley (near Gatwick)	15	15
Bracknell (Berkshire)	20	10
Luton (East Hyde) — Summer	7	7
— Winter	10	10
Basildon U.D.C. (Essex)	15	15
West Hertfordshire (Maple Lodge)	20	15
Bournemouth (Holdenhurst)*	15	20
Swindon (Rodbourne)	10	10
Driffield (East Riding of Yorkshire)	10	20
Dunstable (Bedfordshire)	15	15
Rye Meads (Middle Lee) — Summer	5	5
— Winter	10	10

* Also ammoniacal N not to exceed 20 mg/l. The treatment (comminution, Simplex Activated Sludge and a nitrifying filter for part of flow) is given to 3 × D.W.F. Flows over 3 × D.W.F. receive tank treatment only; suspended solids then must not exceed 150 mg/l.

Where a River Authority requires from a local authority standards for a sewage effluent stricter than those of the Royal Commission, the River Authority must give good reasons for such standards, which must be related to the characteristics of the river into which the effluent discharges and not merely to dilution.

TREATMENT AND DISPOSAL OF SEWAGE IN TIDAL WATERS[10]

Most coastal towns in Britain, and indeed some situated on tidal estuaries, discharge virtually untreated sewage into the sea or estuary. Some of the older outfalls are badly sited and are carrying very much larger volumes of sewage today owing to the increased use of water, increases in resident population and the great increase in visitors to coastal resorts in the summer months. Since the specific gravity of sea water is about 1·026 whereas that of sewage is approximately 1·000, the discharge of untreated sewage from a submerged outlet results in the lighter sewage floating on top as an objectionable greasy film or 'slick' for a considerable distance. Also many of the beaches are aesthetically revolting owing to tides and onshore winds bringing back excreta, paper, etc. Despite, therefore, the Medical Research Council's Report[11] on sewage contamination of bathing beaches denying the risk of bathing in sewage polluted seawater from the health standpoint, public opinion favours some form of treatment, or else a very long outfall. During the past few years, therefore, many coastal towns have been putting their house in order and are either constructing very long outfalls to a point of no return to the shore, or are giving partial or even complete treatment to their sewage. The length of the outfall, the method adopted and the degree of treatment depends upon local conditions, such as the strength of the tidal and other currents, the prevailing winds, the physical characteristics of the foreshore and coastline, the presence of jetties, harbour walls, etc., the depth of the water at different states of the tide, and the proximity of bathing beaches, oyster beds, etc. The direction of the currents can be determined by plotting the course of weighted floats, or by the use of dyes, radioactive tracers or bacterial tracers.

The example set by the seaside town of Bournemouth (cf. *Table 19*) should stimulate other coastal resorts to carry out similar schemes and clean up their polluted beaches. At Bournemouth[12] the comminuted sewage (flows up to $3 \times$ D.W.F.) is settled and treated by the activated sludge Simplex process. A proportion of the flow is treated in a nitrifying filter to give an effluent with an ammoniacal nitrogen content not exceeding 20 mg/l. The final effluent discharges to the River Stover. Sewage from existing sea outfalls will eventually be diverted to the new works. Flows over $3 \times$ D.W.F. receive tank treatment (standard-suspended solids not to exceed 150 mg/l).

Methods available for the treatment of the sewage are as follows:
1. Screening to remove the grosser solids.
2. Maceration by a comminutor to break up the solids into fine particles. This does not, of course, remove the pollution but merely 'disguises' it.
3. Sedimentation to remove settleable solids and so avoid formation of putrefying sludge and discolouration of the sands.
4. Chlorination of the settled sewage. This may be necessary in the summer to prevent bad odours.
5. Aerobic biological treatment (cf. Chapter 6) of the settled sew-

age where a high quality effluent is required, as at Southport and Bournemouth.
6. Any sludges produced should be treated by standard methods.
7. The electrolytic treatment of sewage by the Føyn or Mendia processes gives a 70–80 per cent reduction in B.O.D. and removes suspended matter. It is particularly suitable on the coast where sea water is available.

It is important to stress that storage tanks of sufficient capacity should be provided to hold the settled sewage or effluent so that it can be discharged on the ebb tide only.

A valuable memorandum on the subject of the discharge of trade and sewage effluents into tidal waters was issued by the former Institute of Sewage Purification[13].

ELECTROLYTIC TREATMENT OF SEWAGE[14]

Although the electrolytic treatment of sewage was tried at the end of the last century to produce chlorine which reduced organic impurities and sterilised the sewage, the process was found to be too costly and not very satisfactory. Recently, however, with improved technical knowledge, there has been a revival of interest in the method, particularly for use at coastal towns where sea water is available.
Føyn method—This process was specially devised by E. Føyn to deal with pollution problems in Oslo fjord, Norway, where the excessive fertilisation or 'eutrophication' of the estuary water by nitrogen and phosphorus salts in biologically treated effluents caused algal growths or 'blooms' which decomposed and caused dissolved oxygen deficiency and bad odours in the water.

The Føyn method depends on the removal of phosphate on flocs of magnesium hydroxide[15] formed in alkaline solution in sea water. The process is operated in an electrolytic cell divided by a porous diaphragm. The graphite anode is surrounded by sea water and the iron cathode by sewage mixed with 10–15 per cent of sea water. The alkali formed at the cathode produces a flocculent precipitate containing $Mg(OH)_2$ and $MgNH_4PO_4$ which adsorbs organic impurities, floats to the top with the evolved hydrogen bubbles and is skimmed off. About 90 per cent of the phosphorus in the sewage is thus removed. Chlorine evolved at the anode serves to sterilise the sewage before discharge to the fjord. The scum removed, which contains phosphate, is used as a fertiliser. It must be remembered, however, that the cost of electricity in Norway is much lower than in Britain. In a more recent development of the method, it was found that the diaphragm could be dispensed with by using horizontal electrodes and making use of density differences between the sea water and the sewage.

A similar process, differing in technical details, is used by Mendia in Italy. A demonstration plant has been built in Portsmouth to test the method[15] where it has been found that about 65 per cent of the B.O.D. and suspended solids are removed as well as much reduction in coliform organisms.

Method used in Guernsey[16]—The method used in the island of Guernsey is much simpler. Hitherto, cesspool liquor and septic tank liquor have been discharged direct to the sea, but now a plant has been designed on the south coast of the island, at Le Creux Mahie, to electrolyse sea water and then mix the electrolysed sea water containing active chlorine with the sewage to achieve sterilisation. After passing through a macerator, the resulting effluent is discharged to the sea.

TIP DRAINAGE

The problem of finding suitable sites for the controlled tipping of domestic and industrial refuse has, in recent years, become very acute in the densely populated parts of the country. Sites alongside watercourses are being used, and if suitable precautionary steps are not taken, this can result in polluting discharges from the tips discharging into watercourses. Tip drainage can vary considerably in strength and character, but it usually has a high ammoniacal nitrogen content and can contain appreciable amounts of iron. Sulphide is sometimes present and small amounts of toxic metals are often found in these liquors. The tips are sometimes used for the disposal of sludges and semi-solid wastes and these can give rise to extremely polluting drainage. With the present trend towards cleaner rivers and the previously mentioned emphasis on the conservation of water resources, the drainage from tips can pose a serious problem, and where tips are sited near watercourses it is becoming increasingly evident that steps have to be taken to prevent pollution taking place. The first point to be considered is how to prevent water passing through the tipped material. The tip should not be sited on land where there are springs. Cut-off drains can be laid to prevent water passing through the site from land at a higher level and any ditches or small watercourses passing through the site can be intercepted. The surfacing of the tip can play its part in preventing rainwater passing through the tip. The next point to be considered is the disposal of any drainage from the tip. This can be collected in a drain laid between the tip and any watercourse in the vicinity and the liquor should, if possible, be discharged into the local authority's sewerage system. Treatment on site is not a very satisfactory proposition, especially as the liquor can contain inhibiting constituents, which is more probable if the tip is used for the disposal of industrial refuse. Discharges of tip drainage to sewers need to be kept under strict surveillance by the sewage works management as they can fluctuate greatly in nature and composition. Once a tip has been established it is extremely difficult to completely control what is put on it and illicit tipping can take place. Cases have been reported of drums of cyanide having been found on tips, and if occurrences of this nature take place there is the possibility of the sewage works being seriously affected.

It should always be remembered that the presence and nature of underground water supplies must always be taken into consideration when deciding on the site for a tip.

SPILLAGES ON HIGHWAYS

The amount of liquid, sludge and semi-solid matter conveyed by road tankers has increased considerably over the past few years, not the least contribution to the increase being the volume of waste liquors and sludges from industrial premises being conveyed to disposal points in this manner. From time to time road tankers are involved in accidents which result in their contents, either wholly or in part, being spilled on to the road. Accidents can also occur involving vehicles carrying dry matter which can have polluting and dangerous characteristics. If these materials find their way into road gullies serious problems can arise. Where an accident occurs in an area where the sewers are on the separate system, gross pollution of watercourses can take place. If the liquid is inflammable a fire hazard can ensue and if the watercourse passes through culverts there is also the risk of explosions. Water abstractors can be seriously affected if not warned in time. Where the sewers are on the combined system, dangerous and sometimes explosive conditions can be set up in the sewers, and the sewage treatment plant can be affected with a resulting deterioration in the quality of the final effluent.

In most parts of the country schemes have been agreed between the local authorities, police and fire authorities and River Authorities for the notification of interested parties when spillages of dangerous and polluting substances on to the roads take place, and also for dealing with the spillages. It is obvious that when such incidents occur, the protection of life and property must have first priority. But as a general rule, and wherever possible, it is desirable that the following procedure be adopted:

(a) Notification of interested parties so that precautionary measures can be implemented
(b) The retention, where possible and feasible, of the spilled material to prevent it gaining access to the road gullies
(c) The efficient disposal of the spilled material in a manner which will cause the least trouble.

The latter point is generally the one most difficult to achieve as few local authorities have the means and equipment readily available during the whole of the 24 hours of the day, seven days per week, to deal with such matters, which is unfortunate as time is a very important factor in dealing with these incidents. This is probably a further argument for the setting up, by local authorities, of an emergency service, available at all times, and equipped to deal with a very wide variety of emergencies, as is already provided by several cities in the U.S.A.

STORM SEWAGE OVERFLOWS AND STORM SEWAGE TREATMENT

The Technical Committee on Storm Overflows (Interim Report, 1963) came to the conclusion that 'We accept that the practice of overflowing storm sewage to watercourses must continue for a long time to come'. The reason is that so many combined systems of sewerage exist and their replacement by separate systems in urban areas would be a tremendous and very expensive undertaking. Nevertheless, many existing overflows are unsatisfactory. The Committee also recommended that, in the design of overflows, a storage chamber to accommodate the first flush of very polluting storm sewage should be provided to minimise the amount of floating and solid particles on the overflowing sewage. Where this is not feasible a higher setting may be necessary. The Committee published its Final Report in 1970 (cf. Chapter 3, p. 46).

An improved storm sewage overflow has been designed by Dobbie and Wielogorski[17]. A 'meandering' channel is used to provide helical flow with an overflow weir placed along the outer wall of a bend (total angle between 60° and 90°) from which the less polluted and lighter liquor will be drawn off, while the more polluted liquor concentrates along the inner wall of the bend.

There is a school of thought which considers that separate storm water tanks should be discarded and replaced by extra primary sedimentation tank capacity: all the sewage entering the works would pass through the primary sedimentation tanks, the storm water separation weir would be sited downstream from the tanks, and the liquor passing over the weir would discharge to the river. If, however, all the sedimentation tanks were used in dry weather there would be no storage capacity to prevent any discharge to the river in a storm of short duration, and no means of retaining the first foul flush at the commencement of any storm of longer duration. Furthermore, the first discharge to the river would not be settled storm water but settled dry-weather crude sewage displaced from the tanks.

If the drive for cleaner rivers is to have the degree of success which is desired, great efforts will have to be made in the coming years to solve the problem posed by storm sewage overflows and to improve the treatment of storm sewage at sewage treatment works. Large amounts of money are being expended to ensure that sewage works are capable of producing effluents of Royal Commission standard or better, and it would be most unfortunate if these efforts to clean up the rivers were to be negatived to some extent by the continuation of unsatisfactory discharges of storm sewage.

SEDIMENTATION TANKS

Much work has been carried out on improving the efficiency of sedimentation tanks, especially on the design of inlets and outlets and of sludge removal mechanism. One result of this work has been the development of the notched weir.

PERCOLATING FILTERS

Condition and performance of percolating filters depend to a marked extent upon the speed of rotation of the distributors, i.e. the length of the dosing period. Experiments by Lumb and Barnes[18] at Halifax using conventional filters showed that distributors rotating slowly, at 1 rev in about 4 min, gave better quality effluents and maintained the filters in a cleaner condition than on faster rotation (1 rev in about 44 sec). They concluded from these and other trials that, with Halifax tank effluent, the optimum dosing cycle was within the range 4–9 min; the improvement in the B.O.D. of the final effluent was about 20–25 per cent. The optimum dosing cycle and the improvement in efficiency would not necessarily be the same at other sewage works, depending upon the nature of the sewage and the characteristics of the particular plant. Similar results were obtained by Tomlinson and Hall[19] with alternating double filters at Minworth, Birmingham. Control of the dosing cycle is now effected at some works by using motor driven distributors which can be rotated at different speeds.

Some modern sewage works are being constructed to operate on the two-stage filter system (not alternating double filtration). The new works being constructed by Andover Borough Council[20] will utilise this principle. The settled sewage will first be treated on circular primary filters. The volumetric loading will be 3 400 l/m^3 day (572 gal/yd^3 day) and the B.O.D. loading 0·71 kg/m^3 day (1·2 lb/yd^3 day). After sedimentation the effluent will receive further treatment on rectangular secondary filters with a volumetric loading of 850 l/m^3 day (143 gal/yd^3 day) and a B.O.D. loading of 0·089 kg/m^3 day (0·15 lb/yd^3 day).

ACTIVATED SLUDGE PLANTS

IMPROVEMENTS IN EFFICIENCY

In recent years, much experimental work has been carried out on the intensification of the mechanical systems of aeration. Improved results have been achieved by increasing the rate of oxygen input into the mixed liquor (the 'oxygenation capacity', expressed as g oxygen/m^3 h of tank volume at 10°C and 101 300 N/m^2 (760 mm barometric pressure) by working with much shallower tanks and, in Simplex plants, by using a new type of high-intensity cone and by increasing its speed of revolution, and in Kessener Plants by using an improved and larger type of rotor. In this way, a Royal Commission effluent is obtained using quite short detention periods; in some cases as low as 1½ h at D.W.F. The tendency today is also to carry larger amounts of activated sludge in the aeration units, and the ratio of the volume of activated sludge returned from the final sedimentation tanks to the volume of settled sewage is in some cases as high as 1·5 : 1.

COARSE BUBBLE AERATION

The use of coarse bubble aeration has come to the fore in recent years. One of the systems using this method of aeration is the INKA Aeration Process[21].

This is a development of the activated sludge process which

originated in Sweden. It differs from the conventional diffused air process in the following novel features:

(a) Coarse air bubble aeration (instead of the customary fine bubble aeration) is used, produced with aerator pipes having 2·5 mm (0·1 in) diam. orifices on the underside.
(b) The aerator pipes are in channels about 2·4–4·9 m (8–16 ft) deep and are immersed to a depth of only about 0·8 m (2·5 ft) instead of being on the tank floor.
(c) These pipes can easily be removed, if necessary, for inspection without emptying the tanks.
(d) The low immersion depth means that comparatively low-pressure air, produced by fans, can be used, so maintenance and running costs are low.
(e) Each aeration channel is divided into two communicating compartments by a wall. Air bubbles produced by the aerator pipes in one compartment cause an upward current in this compartment and a downward current in the other compartment, thus producing a clockwise flow which maintains the turbulence necessary to keep the activated sludge in circulation (see *Figure 27*).
(f) Fine air filters are unnecessary.

Figure 27. Basic principle of INKA aeration process. (By courtesy of The Dorr-Oliver Co., Ltd., Croydon)

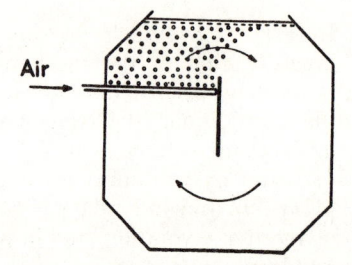

PASSAVANT MAMMOTH ROTOR

This has been recently introduced into this country by Whitehead & Poole Ltd. The rotors, similar in some respects to the Kessener brushes, are 1 m diam. and are fitted in an oval shaped tank with a centre wall in which the mixed liquor is continually circulating. This provides a mixed aeration system.

SLUDGE TREATMENT AND DISPOSAL

The present trend towards mechanisation and automation at sewage works is likely to have repercussions in the field of sludge treatment and disposal. Several new developments along these lines are of interest.

ATOMISED SUSPENSION TECHNIQUE[22]

The crude sludge, concentrated as much as possible by settlement, is disintegrated in a special machine and then atomised with pressure nozzles at the top of a reactor tower, the walls of which are electrically heated (540–760°C). The atomised spray quickly dries and falls down the tower. Air admitted halfway down the tower causes combustion of the organic matter to a harmless ash. The whole process takes only 15 s and is claimed to be superior to conventional methods of treatment and disposal. The technique can also be used as a drying process only.

ZIMMERMANN PROCESS[23]

This is a wet combustion process of the organic matter of sludge at very high pressure ($10 \cdot 13 \times 10^6$ N/m^2 or 100 atm) and rather high temperature (about 240°C) in the presence of air in a special stainless steel reactor. The organic matter is oxidised to carbon dioxide; the effluent, containing chiefly mineral matter, is withdrawn continuously and can be lagooned or used as a fill for land. The process is expensive but will be of use in certain circumstances. It is used to treat sewage sludge at Chicago, U.S.A.

SONIC SCREENS[24]

This method of dewatering sludge, originating at the Wuppertal-Buchenhofen Sewage Works, Germany, consists in filtration of the sludge through special vibrating fine screens and filters of high frequency, which renders the machine very noisy. A sludge of about 75 to 80 per cent moisture content is obtained, i.e. roughly the same as from a vacuum filter. The process is undergoing tests at works in Britain (e.g. Bracknell, Berkshire).

DORR-OLIVER FLUO-SOLIDS (FS) SYSTEM[25]

This is an automatic mechanical and thermal oxidation process for dealing with all types of sludge (including industrial sludges) to give an inert ash. Briefly, the thickened sludge is conditioned with chemicals and dewatered on a vacuum drum filter (provided with interchangeable means of discharge). The resulting sludge cake is then burned in a reactor (temperature 705–815°C, pressure 145×10^3

N/m^2, 21 lb/in^2) in an upward-moving stream of air, thus producing a 'fluidised mixture' of gas and fine particles. The process is somewhat expensive but does give complete sludge treatment.

SLUDGE LIFTING

In the field of sludge disposal, machinery has also been developed for mechanical removal of dewatered sludge from drying beds, usually as a type of dredger mounted on an arm or bridge which spans the drying bed. The object is to speed up removal from the beds and to reduce labour costs. An example is the Templewood-Hawksley 'Sludgemaster' used at Mogden[26].

DENSIFICATION OF ACTIVATED SLUDGE

The satisfactory densification of activated sludge is difficult because it does not settle easily when the moisture content is about 99·5 per cent and because of the long time taken for densification it is apt to produce foul odours. In this connection, a new flotation process of densification practised at Kew is of interest[27, 27a]. Air bubbles are used to cause the sludge to rise to the top and a skimming mechanism removes the floating thickened sludge. In this way, at Kew, a densified activated sludge contaiing up to about 97 per cent moisture (2·6–3 per cent solids) has been obtained with a comparatively low specific resistance.

THE SLUDGE PROBLEM IN THE FUTURE

There is no doubt that sludge treatment and disposal is today one of the sewage works manager's greatest problems, and it does appear that the problem may well be accentuated in the future. The reception of ever increasing quantities of trade effluent into the public sewers is bound to increase the amount of sludge to be dealt with. The use of waste disposal units in the kitchens of domestic dwellings could increase considerably in the future, as could their use in hospitals and other public undertakings. This would tend to transfer some of the problem of disposing of solid domestic refuse from the dustbin to the sewage works where additional amounts of sludge would have to be dealt with, and with the additional problem that there may be changes in the character of the sludge.

Apart from the difficulty of dewatering increasing quantities of sludge there is the ever increasing problem of disposing of the dewatered sludge. Incineration is being given serious consideration by some local authorities and considering the fact that incineration is also becoming an attractive method of disposing of solid domestic refuse, it is probably that in the future incinerators will be used for the dual purpose of burning both sewage sludge and domestic refuse. Sites adjacent to sewage works may be found to be the most suitable for this purpose for several reasons. It may be that the flue gases from these incinerators have to be scrubbed and it could be desirable for these liquors, together with any surplus quenching water, to be conveyed to the sewage works via a separate sewer, if they contain sulphides or other noxious constituents, and it would be desirable for these discharges to be under the control of the sewage works manager. Any waste heat from the incinerator plant could be used with advantage at many sewage works, especially where sludge digestion, heat treatment of sludge or sludge drying is practised.

TRADE WASTES

Great and rapid changes are taking place in the types of processes used by manufacturers, with a resulting change in the character of waste liquors discharging from industrial premises. The sewage works manager needs to keep a vigilant eye on trade waste discharging into the sewers, as some such changes may have far-reaching effects on the operation of the sewage works, and thus frequent inspection of trade waste discharges is an absolute necessity. Cotton bleaching is a case in point of the changes which are taking place. The old traditional method of alkaline boil in kiers, followed by hypochlorite bleach, is being replaced in many works by bleaching with hydrogen peroxide or sodium chlorite, which produces waste liquors with different characteristics.

A variety of organic substances which can have disastrous effects on the biological processes at sewage works, can today be found in some trade effluent discharges e.g. pentachlorophenol can have serious effects on sludge digestion processes and allyl thiourea has a strong inhibiting effect on nitrifying organisms.

The Public Health Act 1961 has allowed local authorities to exercise greater control over those discharges of trade effluent which commenced to be discharged to the sewers prior to 1937, and also to levy charges for the reception and treatment of these discharges.

Biological methods of treatment are finding increasing use in the treatment of trade wastes. Thus, at Great Ryburgh, Norfolk, the first trade waste purification plant of its kind has been dealing with malthouse wastes using a Pasveer oxidation ditch. This is an oval-shaped ditch in which a mixture of the waste and activated sludge is impelled at about 0·3 m/s (1 ft/s) and aerated by rotors similar to the Kessener Brush. The B.O.D. was reduced in this way from about 1 000 mg/l to about 40 mg/l. With further extensions, an even better effluent is expected[28].

Poultry evisceration wastes, after screening and separation of fat, have also been treated in British Columbia by a similar biolgical process, followed up with additional treatment in aerated lagoons, flocculation with alum and final settlement, thus giving an overall B.O.D. reduction of about 93–99 per cent, and a final effluent of about 2–55 mg/l[29]

RADIOACTIVITY

The development of nuclear power and the growing use of radio-isotopes at research establishments, hospitals and industrial premises are likely to continue increasing and so we can expect not only discharges to the sewers of wastes containing much higher levels of radioactivity but also greater numbers of discharges. The Radioactive Substances Act, 1960, mentioned on p. 164 regulates the use of radioactive materials and ensures the safe disposal of radioactive wastes by establishing government control over discharge and disposal of such wastes. Where wastes cannot be dealt with locally, a national disposal service will be available to deal with the wastes. In general there should be sufficient flow in sewerage systems to dilute radioactive effluents to harmless levels of activity.

An interesting development which promises to increase our understanding of some sewage treatment processes is the use of radio-isotopes as 'tracers' for flow measurements and for determining detention periods in sedimentation tanks, aeration tanks and percolating filters[30] (see Chapter 9). The ideal isotope for such experiments is one which emits β and γ radiation, is not easily removed from water by adsorption, has a fairly (but not too) short half-life so that it decays quickly, and is neither too costly nor too dangerous. Bromine-82 (half-life 36 h) used as ammonium bromide (NH_4Br^{82}) fulfils these requirements and has been much used in this type of work. The information obtained by this technique will no doubt lead to many improvements in tank design.

SYNTHETIC DETERGENTS

Since the World War II, synthetic detergents have tended to replace ordinary soap made from animal and vegetable fats and oils. The synthetic detergents (often called syndets for short) are made on a large scale from cheap materials derived from petroleum. They are generally divided into three classes:

1. ANIONIC DETERGENTS

These ionise in water giving an active anion which is responsible for detergent action.
　Examples are:
　(a) Primary and secondary alkyl sulphates

$R-CH_2-OSO_3'$ .... Na^+ 　　　Primary

$\begin{array}{c}R_1\\R_2\end{array}\!\!>\!CH-OSO_3'$ .... Na^+ 　　　Secondary
　　　　　　　　　　　　　　　　　　　(e.g. Teepol)

R, R_1 or R_2 are long hydrocarbon chains or alkyl groups containing 9–15 carbon atoms.
　(b) Sodium alkyl benzene sulphonates

　　These were the most important detergents used before 1965; they are cheap and excellent detergents.
　　General formula: $R-C_6H_4-SO_3'$ .... Na^+

R = alkyl group containing 9–15 carbon atoms.
　A detergent of this type much used before 1965 was sodium tetrapropylene benzene sulphonate whose formula is:

$$C_{12}H_{25}-C_6H_4-SO_3Na$$

It is a mixture of isomers containing branched alkyl groups.
　In a well-purified activated sludge effluent, according to Waldmeyer[31] the average anionic syndet concentration is usually below 1·5 mg/l and many are below 1·0 mg/l. This compares with about 4 mg/l before the introduction of soft detergents.

2. NON-IONIC DETERGENTS

These do not ionise in aqueous solution. The most generally used non-ionic detergents in industry are ethyl oxylated alkyl phenol of the general formula:

$$R.C_6H_4.(OCH_2.CH_2)_nOH$$

where R = an alkyl group with 8 or more carbon atoms and n lies between 8 and 11[32].

Usage of these compounds is at present comparatively small, and only about 0·2–0·6 mg/l are usually present in sewage effluents. Nevertheless, their usage is tending to increase, and it has been shown that non-ionic and anionic syndets yield much more foam when mixed together than when used separately.

These non-ionic detergents are comparatively resistant to breakdown by the activated sludge process or in percolating filters. They are used in many industries, e.g. textile finishing, dyeing and printing, paper manufacture, leather manufacture, synthetic fibre manufactor, coal mining, etc. At the Beckton Works of the Greater London Council, the non-ionic detergent concentration varied between 0·02 and 0·5 mg/l, the average being 0·4 mg/l[31].

3. CATIONIC DETERGENTS

Here the cation is the active part of the molecule. They are usually quaternary ammonium or pyridinium salts, e.g.

$$\begin{matrix}R_1\\R_2\\R_3\\R_4\end{matrix}\!\!>\!\!N^+ \ldots\ldots X' \qquad C_{16}H_{33}-N^+\!\!\!<\!\!\!\bigcirc \ldots\ldots Br'$$

They have strong bacterial properties and are used to some extent in washing utensils and equipment in hotels and restaurants and as sterilising agents in hospitals, but are too expensive for general use.

So, synthetic detergents are really soapless washing or cleansing agents derived from synthetic organic chemicals.

The commercial preparations contain only a relatively small proportion of the active synthetic detergent (about 15–35 per cent). The remainder is made up of 'builders' (e.g. various phosphates, sodium perborate and sodium silicate), carboxymethyl cellulose (which prevents dirt from re-depositing during laundering) and substances which increase foam (e.g. alkanolamides). Over 80 per cent of these detergents are used for domestic purposes, but they also find uses for special purposes in various industries such as, leather, textile, cosmetic, bottle washing, and the de-greasing of metals. The growth of the synthetic detergent industry is well illustrated by the following figures: 13 200 tonnes (13 000 tons) in 1949, and 56 900 tonnes (56 000 tons) in 1964.

Over 90 per cent of the detergents used are of the anionic type, somewhat less that 10 per cent of the non-ionic type and only a small proportion of the cationic type (since these are expensive).

The popularity of synthetic detergents is due to:

1. the intensive advertising campaign and high-pressure salesmanship of the manufacturers.
2. Their effectiveness as washing agents in hard water areas. They are unaffected by hard water as their calcium salts are soluble in water whereas calcium salts of ordinary soaps are insoluble and form a scum on the surface of the water.

Up to about 1964 the detergents mainly used for domestic purposes were of the alkyl benzene sulphonate type (ABS), i.e.

$$R-C_6H_4-SO_3' \ldots \ldots \ldots \ldots \ldots Na^+$$

where $R =$ an alkyl group containing 9–15 carbon atoms, the best detergents being those with an alkyl group having 12 carbon atoms. A detergent of this type much used in the past was sodium tetrapropylene benzene sulphonate

$$C_{12}H_{25}-C_6H_4-SO_3Na$$

Unfortunately, this particular detergent, in common with other alkyl benzene sulphonates, had adverse effects at sewage works and in rivers. It caused tremendous amounts of foam (even as little as 0·7 mg/l, expressed in terms of Manoxol O.T., can cause foam). The most affected of all are the diffused air activated sludge plants, where foam can be blown into the air and cause complaints. Syndets also lower the efficiency of the activated sludge process by reducing the transfer of oxygen from the air to the sewage undergoing aeration. The foam is potentially hazardous not only because of its high concentration of bacteria (some of which may be pathogenic) but also on account of its bad effect on crops. Various remedies have been used to suppresses this foam, e.g. spraying with final effluent (as at Nottingham), or the use of anti-foam chemicals (as at Mogden and Davyhulme).

The experimental work of Hammerton[33] and other workers showed that the resistance of anionic detergents to biological breakdown was due to the structure of the alkyl group present in the molecule. Detergents such as Teepol or Drene containing a straight chain alkyl group (e.g. $CH_3(CH_2)_n-$) are easily oxidised biochemically but those containing a branched chain alkyl group, especially quaternary carbon atoms, are most resistant to oxidation. The most widely used detergent hitherto had been sodium tetrapropylene benzene sulphonate (also called sodium Dobane PT sulphonate) which is a mixture of several isomers including the following:

$$CH_3-CH_2-CH_2-CH_2-\overset{CH_3}{\underset{CH_3}{*C}}-CH_2-CH_2-\overset{CH_3}{\underset{CH_3}{*C}}-\langle\rangle SO_3Na$$

$$\overset{CH_3}{\underset{CH_3}{CH}}-CH_2-\overset{CH_3}{\underset{CH_3}{*C}}-CH_2-CH_2-\overset{CH_3}{\underset{CH_3}{*C}}-\langle\rangle SO_3Na$$

It will be noticed that the alkyl groups are much branched (the starred quaternary carbon atoms), being very difficult to oxidise biochemically. On the other hand, straight chain detergents such as sodium n-dodecyl-p-benzene sulphonate, $CH_3(CH_2)_{11}\langle\rangle SO_3Na$, are almost completely oxidised biochemically, but these would be too expensive to use on a large scale. The Technical Committee on Synthetic Detergents[32, 34] recommended experiments to be carried out on a large scale and this was done at the Luton sewage works where it was shown that whereas only about 67 per cent of the old detergent was removed biochemically, about 94 per cent of a new detergent could be removed. It was at this stage that the terms 'hard' detergent (i.e. resistant to biological oxidation) and 'soft' detergent (i.e. relatively *not* resistant to biological oxidation) were coined.

The Americans refer to the older hard alkyl benzene sulphonates as 'ABS', whilst the newer soft detergents are called 'LAS' (linear alkyl sulphonates). The Committee recommended that at the end of 1964, the manufacturers should cease to use the older harder material for anionic detergents[32]. Before the changeover to soft detergents the degree of removal at sewage works was of the order of 68 per cent or less, depending on the efficiency of the works, but with the newer soft detergents over 90 per cent of the detergent is oxidised biologically. In practice, this means that the concentration of detergents in the final effluent instead of being about 32 per cent of the detergent reaching the works, is now about 10 per cent or less. Thus there is much less foam on the rivers into which the sewage effluents are discharged, particularly those derived from sewages containing relatively small volumes of trade effluent†. At the Maple Lodge Works of the West

† Some rivers in industrial areas still cause nuisance by foam. This may be due to the use of the older hard detergents by certain industries, or to kier liquors and other waste waters containing detergents.

Hertfordshire Main Drainage Authority the removal of syndets was 58 per cent in 1955 and 94 per cent in 1966. According to the Ministry of Technology, the percentages removed of some of the newer soft detergents now being used are as follows:

Sodium Dobane JN sulphate (used in Luton experiment)—80 per cent
Sodium Dobane JNX benzene sulphonate—91 per cent
Sodium Dobane JN sulphonate 036—93 per cent

Further research is being carried out to find even softer detergents. As a result, there has been a marked reduction in detergent concentration in many rivers.

Unfortunately, the newer soft detergents, in large quantities, affect anaerobic digestion of sewage sludge adversely, just as do the hard detergents[35]. Inhibition of sludge digestion does not occur until the detergent concentration is about 1·5 per cent, calculated on the dry sludge solids (i.e. rather more than about 750 mg/l in the wet sludge). At a concentration above this, there is increasing reduction in the rate of gas production, and above 2 per cent (calculated on the dry matter) marked inhibition of digestion occurs.

The national average concentration of synthetic detergent in sewage is about 18 mg/l, but when the concentration of synthetic detergent (hard or soft) in settled sewage is higher then about 25–30 mg/l, some inhibition of digestion occurs, leading to a reduction in gas production. This has already happened at some sewage works in the south of England, where, owing to the hard water, there is a greater domestic consumption of synthetic detergents. Studies are being made at the Water Pollution Research Laboratory to deal with this matter. The addition of commercial octadecylamine ('stearine amine')[31] has been used to counteract the adverse effects of anionic detergents on sludge digestion.

A standard test has been devised by the Standing Technical Committee on Synthetic Detergents[36] for assessing the biodegradability of an anionic detergent. The principle of the test is to incubate at 20°C the detergent in a standard synthetic water containing mineral nutrients with a seed sludge from an activated sludge plant. The concentration of detergent is then determined at intervals by the standard methylene blue method and a graph is constructed of per cent detergent remaining versus time of incubation in days.

It may be added that the removal of detergents from sewage is mainly by biodegradation, either by percolating filters or in the activated sludge process. Nevertheless, there is some absorption of the detergent by the primary sludge, usually about 10–30 per cent.

A new series of biodegradable non-ionic detergents has now been manufactured, e.g. Tergitol non-ionic 15 – S – 9

$$CH_3 - (CH_2)_n - CH_3$$
$$|$$
$$O - (CH_2 - CH_2O)_9 H$$

The differentiation between soft detergents ('LAS') and hard detergents ('ABS') in waters can be carried out after activated carbon treatment and elutriation with water by measuring their infra-red absorption spectra at 7·1 μm for LAS and 7·3 μ m for ABS[37]. In this way, the proportions of the two types of detergent can be determined to within 10 per cent.

REFERENCES

1. *Ministry of Technology. Notes on water pollution,* No. 31, 1965, *Reclamation of water from sewage effluents*
2. EDEN, G. E., TRUESDALE, G. A., WYATT, K. L. and STENNETT, G. V., Water from sewage effluents, *J. Proc. Inst. Sew. Purif.* 5 (1966) 407; also *Chem. Ind.* (1966) 1517; Notes on water pollution, No. 31, Dec. 1965; *Instn. Publ. Hlth. Engrs.* 67 (1968) 75
3. *Ministry of Technology. Water Pollution Research,* 1966. p. 120, 1967. London; H.M.S.O.; see also *J. Instn. Publ. Hlth Engrs* 67 (1968) 75
4. STANBRIDGE, H. H., Load equalisation as applied to sewage treatment, *J. Proc. Inst. Sew. Purif.* 4 (1956) 411
5. ANON., Solving the problem of odour control, *Wat. Waste Treat J.* 10 (1965) 525
6. SANTRY, I. W., Hydrogen sulphide odour control measures, *J. Wat. Pollut. Control Fed.* 38 (1966) 459
7. *Ministry of Housing and Local Government. Technical problems of River Authorities and Sewage Disposal Authorities in laying down and complying with limits of quality for effluents more restrictive than those of the Royal Commission,* 1966. London; H.M.S.O.
8. *Ministry of Housing and Local Government. Sewage Effluents,* Circular No. 37/66, 1966. London; H.M.S.O.
9. BOLTON, R. L. and KLEIN, L., Better sewage effluents—their need and their attainment, *Publ. Wks. New Jersey* 93 (1962) No. 10. 109
10. WEBBER, N. B., The discharge of sewage into the sea, *Surveyor, Lond.* 119 (1960) 1421
11. *Medical Research Council* Memo. No. 37, *Sewage contamination of bathing beaches in England and Wales,* 1959. London; H.M.S.O.
12. ANON., Opening of Bournemouth Purification Works, *Surveyor, Lond.* 125 (1965) No. 3800, 55
13. Institute of Sewage Purification, Memorandum on discharge of sewage and trade effluents into tidal waters. *J. Proc. Inst. Sew. Purif.* 2 (1964) 119
14. MARSON, H. W., Electrolytic methods in modern sewage treatment: a summary, *Effluent and Water Treatment J.* 7 (1967) 70, 75
15. ESCRITT, L. B., Electrolytic treatment of sewage is revived at Oslo, *Munic. Engng, Lond.* 137 (1960) 1926
16. MARSON, H. W., Electrolytic sewage treatment: the modern process, *Wat. Pollut. Control* 66 (1967) 109
17. DOBBIE, C. H. and WIELOGORSKI, J. W., Storm water separation by helical motion, *Surveyor. Lond.* 127 (1966) No. 3839, 9
18. LUMB, C. and BARNES, J. P., The periodicity of dosing percolating filters, *J. Proc. Inst. Sew. Purif.* 1 (1948) 83
19. TOMLINSON, T. G. and HALL, H., The effect of periodicity of dosing on the efficiency of percolating filters, *J. Proc. Inst. Sew. Purif.* 1 (1955) 40
20. ANON., Dual purpose settlement tanks at Andover, *Surveyor, Lond.* 133 (1969) No. 4013, 29
21. GEIGER, H., An aeration process with a small injection depth, *Gesundheitsingenieur* 79 (1958) 13
22. ANON., Atomisation technique for sludges, *Wat. Waste Treat. J.* 7 (1960) 492

23. ANON., Sludge burns under water in new incinerator, *Engng News Rec.* 161 No. 22 (1958) 44; see also ZIMMERMANN, F. J., (1954) *Brit. Patent No.* 706,686
24. KIESS, F. and SCHRECKEGAST, C., Sludge dewatering by vibrating screens, *Wat. Sewage Wks.* 106 (1959) 479
25. ANON., *Surveyor, Lond.* 123 (1964) No. 3750, 48; also Dorr-Oliver F.S. Disposal system (1963) Bull. No. 6055, Dorr-Oliver Co. Inc., U.S.A.
26. TOWNEND, C. B., Recent Middlesex developments in mechanisation and automation of sewage plant operation, *J. Proc. Inst. Sew. Purif.* 4 (1961) 273
27. WRIGLEY, K. J., Kew sewage treatment works: operation and performance, *Wat. Pollut. Control* 66 (1967) 247
27a. BROWN, P. and THOMAS, A., Some experience in the consolidation of activated sludge, *Wat. Pollut. Control* 68 (1969) 203
28. ANON., Pasveer ditch applied to a trade effluent, *Surveyor, Lond.* 127 (1966) No. 3844, 39
29. KILBURN, D. G. and TRUSSEL, P. C., A system for the treatment of poultry plant wastes, *Wat. Waste Treat J.* 10 (1966) 618
30. EDEN, G. E., Some uses of radioisotopes in the study of sewage treatment processes, *J. Proc. Inst. Sew. Purif.* 4 (1959) 522
31. Symposium—Pollution by synthetic detergents, *Wat. Pollut. Control* 67 (1968) 56–123
32. Ministry of Housing and Local Government, 9 Progress Reports of the *Standing Technical Committee on Synthetic Detergents*, 1959–1967. London; H.M.S.O.
33. HAMMERTON, C., Observations on the decay of synthetic detergents in natural waters, *J. Appl. Chem., Lond.* 5 (1955) 517; see also *J. Proc. Inst. Sew. Purif.* 3 (1957) 280
34. Ministry of Housing and Local Government, *Report of the Committee on Synthetic Detergents*, 1956. London; H.M.S.O.
35. BRUCE, A. M., SWANWICK, J. D. and OWNSWORTH, R. A., Synthetic detergents and sludge digestion: some recent observations. *J. Proc. Inst. Sew. Purif.* 5 (1966) 427
36. Ministry of Housing and Local Government, *Supplement to the 8th progress report of the Standing Technical Committee on Synthetic Detergents*, 1966. London; H.M.S.O.
37. MAEMLER, C. Z., CRIPPS, J. M. and GREENBERG, A. E., Differentiation of LAS and ABS in water. *J. Wat. Pollut. Control Fed.*, Research Supplement, R92–R98, 39 (1967)

Chapter 13

CHEMICAL CALCULATIONS

CALCULATIONS INVOLVING PER CENT PURIFICATION OF SEWAGE

If D_s = chemical determination (e.g. B.O.D., suspended solids, etc.) for raw sewage in mg/l

and D_E = corresponding determination for final effluent in mg/l

then per cent purification (or removal of constituent)

$$= \frac{100 (D_s - D_E)}{D_s}$$

Thus to take three examples from the analyses given in *Table 9*:

1. Removal of suspended solids in settling tanks $= \dfrac{100 \ (157 - 35)}{157}$

   $= 78$ per cent

2. Overall B.O.D. reduction $= \dfrac{100 \ (347 - 12 \cdot 4)}{347}$

   $= 96$ per cent

3. Overall 4 h permanganate value reduction

   $= \dfrac{100 \ (208 - 22 \cdot 4)}{208}$

   $= 89$ per cent

CALCULATIONS IN VOLUMETRIC ANALYSIS

In general volumetric analysis, solutions of known concentration are used, referred to as 'standard solutions'. These are usually based on the concept of 'Normality'. A 'Normal' solution of a substance contains its 'equivalent', i.e. the weight of substance in grammes which, *in the reaction under consideration*, combines with or displaces 1

Chemical Calculations

gram-atom (the atomic weight in grams) of hydrogen (1·008 g). A normal solution (denoted by the capital letter N), thus contains the equivalent-weight of the substance per litre of solution. The equivalent is the weight of substance in grammes which, *in the particular reaction under consideration*, combines with or displaces 1 gram-atom of hydrogen (1·008 g), or 1 gram-atom of hydroxyl (17·008 g), or 1 gram-atom of any monovalent ion (e.g. 127 g of iodine), or half a gram-atom of a divalent ion such as oxygen (8 g).

In general, if y molecules of a substance react with, displace, or yield z atoms of a monovalent ion and the molecular weight of the substance is M (= sum of atomic weights of the elements present), then Equivalent weight = $\dfrac{M \times y}{z}$

If the atom is divalent instead of monovalent (e.g. oxygen), it must be remembered that

$$1 \text{ O atom} = 2 \text{ H atoms}$$

Thus, in oxidation-reduction reactions, potassium permanganate in dilute sulphuric acid solution, can supply 5 oxygen atoms for oxidation purposes:

$$2KMnO_4 \rightarrow K_2O + 2MnO + 5O$$

(in sulphuric acid solution these oxides are converted to sulphates)
Since $5O = 10H$ the equivalent weight in this particular reaction

$$= \frac{M \times 2}{10} = \frac{1}{5} \text{ of } 158 = 31 \cdot 6$$

Hence, a normal solution contains 31·6 g of potassium permanganate per litre of solution.

In the determination of dissolved oxygen, or of permanganate value, the iodine liberated from the potassium iodide is titrated with a standard volume of hydrated sodium thiosulphate, the reaction being:

$$Na_2S_2O_3.5H_2O + I_2 = Na_2S_4O_6 + 5H_2O + 2I^-$$
$$(248 \cdot 19)$$

Thus, 248·19 g of hydrated $Na_2S_2O_3$ react with 2 iodine atoms (equivalent to 1 oxygen atom since O is divalent).

Hence 248·19 g of $Na_2S_2O_3.5H_2O$ are equivalent to 8 g oxygen or 3·1024 g of $Na_2S_2O_3.5H_2O$ are equivalent to 0·1 g oxygen.

Another and more modern way of looking at equivalent weights in oxidation reactions is to determine the number of electrons transferred. Oxidation involves a loss of electrons whilst reduction involves a gain of electrons. If M is the molecular weight, and n the number of electrons (e), gained or lost by one molecule of the substance, then:

$$\text{Equivalent weight} = \frac{M}{n}$$

Examples

$$Fe^{++} \leftrightarrows Fe^{+++}$$

Equivalent weight of Fe = 55·85

In the reactions involving the oxidation of ferrous iron to ferric iron, we have:

$$MnO_4' + 8H^+ + 5e \rightarrow Mn^{++}$$

So, equivalent weight of $KMnO_4 = \dfrac{KMnO_4}{5} = 31\cdot 6$

$$Cr_2O_7'' + 14H^+ + 6e \rightarrow 2Cr^{+++}$$

So, equivalent weight of $K_2Cr_2O_7 = \dfrac{K_2Cr_2O_7}{6} = 49\cdot 04$

CALCULATIONS IN VOLUMETRIC ANALYSIS

In many technical analyses, and this includes sewage, trade wastes and river water, calculations are much simplified by using standard

Table 20. SOME STANDARD SOLUTIONS USED IN THE VOLUMETRIC ANALYSIS OF WASTE WATERS

Constituent to to be determined	Standard solution (s)	Formula of standard substance	Concentration of standard solution (g/l of solution)	1 ml of standard solution is equivalent to
Dissolved Oxygen	Sodium thiosulphate	$Na_2S_2O_3 \cdot 5H_2O$	3·1024†	0·1 mg of oxygen
Permanganate value	Sodium thiosulphate	$Na_2S_2O_3 \cdot 5H_2O$	3·1024†	0·1 mg of oxygen
Chloride	Silver nitrate	$AgNO_3$	4·791*	1 mg of Cl'
Nitrate (titration method)	Sulphuric acid, Sodium hydroxide	H_2SO_4 NaOH	N/140‡ N/140‡	0·1 mg of N
Ferrous iron	Potassium dichromate	$K_2Cr_2O_7$	0·8782* 8·782*	1 mg of Fe" 10 mg of Fe"

† Should be standardised against a primary standard and then adjusted to exactly N/80 strength.
* Can be prepared by direct weighing of the standard substance as these reagents are generally of high purity.
‡ Prepared by special standardisation methods.

Chemical Calculations

solutions of such a strength that 1 ml is equivalent to exactly 10 mg, 1 mg or 0·1 mg of the constituent to be determined. Some examples are shown in *Table 20*.

SLUDGE: WEIGHT–VOLUME–MOISTURE–SOLIDS RELATIONSHIPS

If water is removed from sludge by settlement, filter-pressing, vacuum-filtration or digestion, the weight and volume alter approximately in accordance with the formulae:

$$\frac{W_1}{W_2} = \frac{V_1}{V_2} = \frac{100 - M_2}{100 - M_1} = \frac{S_2}{S_1}$$

where W_1 and W_2 are the sludge weights
V_1 and V_2 are the sludge volumes
S_1 and S_2 are the dry solids contents (per cent by weight)
and M_1 and M_2 are the moisture contents (per cent by weight)

These formulae are not absolutely correct since the removal of water from sludge alters its specific gravity slightly, but for all practical purposes they are accurate enough.

It will be seen that the weight or volume of a sludge varies inversely as the percentage of dry solids by weight.

Example 1
What is the volume of 1 000 gal of 95 per cent moisture sludge if the moisture content is reduced to 80 per cent?

$$\frac{V_1}{V_2} = \frac{1\,000}{V_2} = \frac{100 - 80}{100 - 95} = \frac{20}{5} = 4$$

i.e. the volume is reduced to 250 gal or to ¼ of its original volume.

The advantage of de-watering sewage sludge as much as possible is obvious, since we reduce the volume to be disposed of.

Example 2
100 tons of wet sludge (92 per cent moisture) are pressed to give a sludge cake containing 56 per cent moisture. What is the weight of sludge cake obtained?

$$\frac{100}{W_2} = \frac{100 - 56}{100 - 92} = \frac{44}{8}$$

$$\text{therefore } W_2 = \frac{100 \times 8}{44} = 18 \cdot 2 \text{ ton}$$

Example 3

A sewage flow averages 5×10^6 gal/day and contains 350 mg/l of suspended solids. The sedimentation tank effluent averages 150 mg/l of suspended solids. If the sedimentation tank sludge has a specific gravity of 1·03 and contains 95 per cent of moisture, what is the daily volume of wet sludge produced?

Dry solids removed = 350 − 150 mg/l = 200 mg/l
= 200 lb/1 000 000 lb of sewage
= 10 000 lb/5 000 000 gal of sewage.

Wet sludge produced, at 95 per cent moisture, contains 5 per cent of dry solids.

$$\text{Therefore, weight of wet sludge} = \frac{100}{5} \times 10\,000 \text{ lb}$$

$$= 200\,000 \text{ lb}$$

1 gal of this sludge will weigh $10 \times 1\cdot03 = 10\cdot3$ lb

$$\text{Therefore, volume of sludge produced} = \frac{200\,000}{10\cdot3} \text{ gal/day}$$

$$= 19\,417 \text{ gal/day}$$

For all practical purposes, this could be taken to be 20 000 gal/day.

REDUCTION OF ORGANIC MATTER BY DIGESTION

The reduction in organic matter during sludge digestion can be calculated by using a formula due to Van Kleeck (see Chapter 8), but the student is advised to work it out from first principles on the assumption that the mineral matter in the sludge remains constant during digestion. This assumption is only approximately true since some mineral matter (probably some 5 per cent) passes into the liquor.

Example

A raw sludge containing 75·4 per cent of dry organic matter gives on digestion a sludge containing 62·4 per cent of dry organic matter.
Calculate the percentage loss of organic matter by digestion.

100 g crude sludge contain 75·4 g dry organic matter and 24·6 g mineral matter.

100 g digested sludge contain 62·4 g dry organic matter and 37·6 g mineral matter

After digestion 37·6 g of mineral matter is associated with 62·4 g dry organic matter.

Chemical Calculations

Therefore, original 24·6 g of mineral matter would be associated with $\dfrac{62\cdot4 \times 24\cdot6}{37\cdot6}$ g dry organic matter

= 40·8 g dry organic matter.

Therefore, loss of dry organic matter

$$= \left(\frac{75\cdot4 - 40\cdot8}{75\cdot4}\right) \times 100 \text{ per cent}$$

= 46 per cent

CALCULATION OF AREA OF MEDIA REQUIRED FOR PERCOLATING FILTERS

A sewage with a flow of 1 900 000 gal/day and a settled B.O.D. of 150 mg/l is to be treated on percolating filters. Assuming a filter depth of 6 ft calculate:
  (i) total area of filter media to produce a Royal Commission effluent.
  (ii) the diameter of the circular filters if 10 beds are to be provided.

Method 1 (Based on hydraulic loading)

The sewage is weak and is, therefore, dosed at the rate of 100 gal/yd³ day.

100 gal/need 1 yd³ of media

Therefore, 1 900 000 gal need $\dfrac{1\ 900\ 000}{100}$ = 19 000 yd³ of media

Filter depth = 6 ft = 2 yd

Therefore, area of media = $\dfrac{19\ 000}{2}$ = 9 500 yd²

Method 2 (Based on B.O.D. loading)

Assume B.O.D. loading of 0·15 lb/yd³ day

150 mg/l B.O.D. = 150 lb B.O.D./100 000 gal sewage

= 2850 lb B.O.D./1 900 000 gal of settled sewage

Therefore, volume of media required = $\dfrac{2850}{0\cdot15}$ = 19 000 yd³

Therefore, area of media = $\dfrac{19\ 000}{2}$ = 9 500 yd²

To calculate the diameter of each filter, assuming 10 filters are used, use the well-known area formula:

$$\pi r^2 = \frac{\pi d^2}{4} \quad \text{(where } d = \text{diameter)}$$

$$= 0.7854 \, d^2$$

For 10 filters the diameter d is calculated as follows:

$$\text{Area of each filter} = \frac{9\,500}{10} = 950 \text{ yd}^2$$

Therefore, $0.7854 \, d^2 = 950 \text{ yd}^2$

$$d^2 = \frac{950}{0.7854} = 1\,209.6$$

Therefore, $d = \sqrt{1\,209.6}$

$$= 34.78 \text{ yd (can be calculated by log tables)}$$

Therefore the diameter of each filter is 34.78 yd.

PHELPS' LAW

According to this Law, the biochemical oxidation of sewage by bacteria proceeds in conformity with the equation:

$$\log 10 \frac{L}{L - x} = kt$$

where L = first stage or carbonaceous oxygen demand (this is usually about 20 days)

x = oxygen demand (in mg/l) in t days, as determined by the B.O.D. test.

k = a constant

t = time (in days)

A problem (cf *J. Proc. Inst. Sew. Purif.* Pt. 3 (1964) 297; details of calculation *not* given) set in the 1963 Associate Membership Examination of the Institute of Sewage Purification (reproduced by courtesy of the Institute) was to calculate the first-stage oxygen demand for a settled sewage having a 5-day B.O.D. of 300 mg/l, given that the value of k for sewage is 0·10 (this value would, of course, be different for other types of organic waste waters).

We have x = 300 mg/l
t = 5 days
k = 0·1

Chemical Calculations

Therefore, $\log_{10} \dfrac{L}{L - 300} = 0{\cdot}1 \times 5 = 0{\cdot}5$

antilog $0{\cdot}5 = 3{\cdot}162$ (from antilogarithm table)

Hence $\dfrac{L}{L - 300} = 3{\cdot}162$

Therefore, $L = 3{\cdot}162\,(L - 300)$

$\qquad\qquad = 3{\cdot}162\,L - 948{\cdot}6$

Therefore, $2{\cdot}162\,L = 948{\cdot}6$

$$L = \dfrac{948{\cdot}6}{2{\cdot}162} = 438{\cdot}8 \text{ mg/l}$$

The answer could well be given as 439 mg/l to the nearest unit.

INTERCONVERSION OF BRITISH AND SI UNITS[1–9]

The system of British units long in use in this country and the British Commonwealth is being replaced gradually by a modified form of the metric system. This modern form of the metric system was agreed internationally in 1960 and termed the '*Système International des Unités*', abbreviated to SI units. A few non-SI units (e.g. the litre) are to be retained. Some of the commoner factors for the interconversion of the British and SI units are given below in *Table 21* (see also References 1–9). But it is important for the student to know how these factors are obtained, so a sample calculation is given:

Example

Find the factor for converting lb B.O.D./1 000 ft^3 day to g B.O.D./m^3 day, (a unit used in activated sludge plant practice).

We have
$$\text{lb} \times 453{\cdot}6 = \text{g}$$
$$\text{ft}^3 \times 0{\cdot}02832 = \text{m}^3$$
$$1\,000\text{ ft}^3 \times 28{\cdot}32 = \text{m}^3$$

Therefore,
$$\dfrac{\text{lb} \times 453{\cdot}6}{1\,000\text{ ft}^3 \times 28{\cdot}32} = \dfrac{\text{g}}{\text{m}^3}$$

$$\dfrac{453{\cdot}6}{28{\cdot}32} = 16$$

The factor for converting lb B.O.D./1 000 ft^3 day to g B.O.D./m^3 day is, therefore, 16. To convert the metric relation back to the British one, divide by 16.

Tables for converting British Units to SI Units and SI Units to British Units can be found in *Table 21* and in Appendix 2.

Table 21. SOME CONVERSION FACTORS: BRITISH UNITS AND METRIC UNITS

To convert A to B, multiply A by the factor.
To convert B to A, multiply B by the reciprocal.

Quantity	British Unit A	SI Unit B	Factor	Reciprocal
Length	in	m	2·54	0·3937
	ft	m	0·3048	3·281
	yd	m	0·9144	1·0936
Area	ft^2	m^2	$9·29 \times 10^{-2}$	10·76
	yd^2	m^2	0·8361	1·196
	acre	hectare	0·4047	2·471
Volume	ft^3	m^3	$2·832 \times 10^{-2}$	35·315
	yd^3	m^3	0·7646	1·308
Capacity	gal	l	4·546	0·22
	gal	m^3	$4·546 \times 10^{-3}$	220
Rate of	10^6 gal/day	m^3/s	$5·262 \times 10^{-2}$	19
flow	ft^3/s	m^3/s	$2·832 \times 10^{-2}$	35·315
Velocity	ft/s	m/s	0·3048	3·281
Concentration	lb/100 000 gal	mg/l	1	1
Pressure	lb force/in^2	N/m^2	$6·894 \times 10^3$	$1·4504 \times 10^{-4}$
Weight	lb	kg	0·4536	2·2046
(mass)	ton	kg	1016	$9·842 \times 10^{-4}$
Power	hp	kW	0·7457	1·341
Energy	Btu	J	1 055·06	$9·4782 \times 10^{-4}$

REFERENCES

1. *Ministry of Technology, Changing to the Metric System, conversion factors, symbols and definitions*, Revised Ed., 1969. London; H.M.S.O.
2. *British Standards 3763: 1964, The International System (SI) units*, 1964. London; British Standards Institution
3. *The use of SI units*. 2nd Ed. PD 5686. 1968. British Standards Institution Sales Office, 101–113, Pentonville Road, London, N.1.
4. BOULTON, A. G., *Metrication*, 1966. Reading; Water Resources Board
5. *Ministry of Housing and Local Government, Metric Units with reference to water, sewage, and related subjects. Report of working party*, 1968. London; H.M.S.O. See also *Wat. Pollut. Control* 67 (1968) 475
6. British Standard 350: Part 2: 1962. *Conversion factors and tables* (gives detailed conversion factors of British units to Metric units and *vice versa*). London; British Standards Institution
7. TEBBUTT, T. H. Y., Metric units in water pollution control. *Wat. Pollut. Control* 69 (1970) 48
8. *Standard Mathematical Tables* (includes decimal conversion tables). 17th Ed., 1970. Oxford; Blackwell Scientific Publications
9. *Symbols, signs and abbreviations*, 1969. London; Royal Society

APPENDIX 1

SUGGESTIONS FOR FURTHER READING

WISDOM, A. S., *The law on the pollution of waters*, 2nd Ed. 1966. London; Shaw

WISDOM, A. S., *The law of rivers and watercourses*, 1962. London; Shaw

BATESON, T. C. H., A student's introduction to sewage works design, 1958. London; Pitman

ESCRITT, L. B., Sewerage and sewage disposal: calculations, design, and specification, 3rd Ed. 1965. London; C. R. Books Ltd.

IMHOFF, K., MÜLLER, W. J. AND THISTLETHWAYTE, D. K. B., *Disposal of sewage and other water-borne wastes*, 2nd Ed. 1971. London; Butterworths

TEMPLE, F. C., *Small sewage works*, 1955. London; Technical Press

MCCABE, J. AND ECKENFELDER, W. W., *Biological treatment of sewage and industrial wastes*. Vol. I. Aerobic Oxidation, 1956. Vol. II. Anaerobic digestion and solids-liquid separation, 1958. New York; Reinhold

WYLIE, J. C., *Fertility from town wastes*, 1955. London; Faber

WYLIE, J. C., *The wastes of civilisation*, 1959. London; Faber

KLEIN, L., *River Pollution: I. Chemical Analysis*, 1959. London; Butterworths

KLEIN, L., *River Pollution: 2. Causes and Effects*, 1962. London; Butterworths

KLEIN, L., *River Pollution: 3. Control*, 1966. London; Butterworths

TAYLOR, E. W., *The examination of waters and water supplies* (Thresh, Beale and Suckling, Ed.), 7th Ed., 1958. London; Churchill

RUDOLFS, W. (Ed.), *Industrial wastes: their disposal and treatment*, 1953. New York; Reinhold

GURNHAM, C. F., *Industrial waste water control*, 1965. New York; Academic Press

KEY, A., *Gas works effluents and ammonia*, 2nd Ed., rev. by GARDINER, P. C., 1956. London; Institution of Gas Engineers

COLLINS, J. C. (Ed.), *Radioactive wastes—their treatment and disposal*, 1960. London; Spon

ISAAC, P. C. G. (Ed.), *Waste treatment*, 1960. London; Pergamon Symposium on Pollution by Synthetic Detergents: Towards a solution, *Wat. Pollut. Control* 67 (1968) pp. 56–123

TINTOMETER LTD., *Colour measurement and public health*, 2nd Ed., 1967. Salisbury

SCHWARZENBACH, G. and FLASCHKA, H., *Complexometric titrations*, 2nd English Ed., 1969. London; Methuen

EWING, G. W., *Instrumental methods of chemical analysis*, 3rd Ed., 1969. London; McGraw Hill

WILSON, E. M., *Engineering Hydrology*, 1969. London; Macmillan

Notes on Water Pollution. A useful series of leaflets published quarterly by the Water Pollution Research Laboratory (formerly under the Department of Scientific and Industrial Research, now under the Ministry of Technology): They are free.

No. 1. *Effect of organic discharge on the level of oxygen in a stream.* June, 1958
No. 2. *Waste waters and fish.* September, 1958
No. 3. *Treatment of electroplating wastes.* December, 1958
No. 4. *Anaerobic digestion of industrial wastes.* March, 1959
No. 5. *Operation of percolating filters.* June, 1959
No. 6. *Discharge of sulphates into concrete sewers.* September, 1959
No. 7. *Continuous recording of dissolved oxygen in rivers.* December, 1959
No. 8. *Sampling natural waters and polluting liquids.* March, 1960
No. 9. *Septic tanks.* June, 1960
No. 10. *Oxygen demand tests for effluents.* September, 1960
No. 11. *Pollution of estuaries:* Part 1, *Effects of pollution.* December, 1960
No. 12. *Pollution of estuaries:* Part 2, *Methods of survey.* March, 1961
No. 13. *Some effects of pollution on fish.* June, 1961
No. 14. *Some recent observations on percolating filters.* September, 1961
No. 15. *Synthetic detergents.* December, 1961
No. 16. *Aeration in the activated sludge process.* March, 1962
No. 17. *Waste waters from farms.* June, 1962
No. 18. *Effects of plants and mud on the level of oxygen in a stream.* September, 1962
No. 19. *Flow measurement by means of tracers.* December, 1962
No. 20. *Some short investigations by the Water Pollution Research Laboratory.* March, 1963
No. 21. *Stratification and failure of gasification in sludge digestion tanks.* June, 1963
No. 22. *"Polishing" of sewage works effluents.* September, 1963
No. 23. *Nitrification in the activated sludge process.* December, 1963

Suggestions for further reading

No. 24. *Some further observations on waste waters from farms.* March, 1964
No. 25. *Measurement of suspended solids.* June, 1964
No. 26. *Recent developments in the determination and recording of dissolved oxygen.* September, 1964
No. 27. *A study of some fishless rivers.* December, 1964
No. 28. *Effects of pollution on the Thames Estuary.* March, 1965
No. 29. *Synthetic detergents and sludge digestion.* June, 1965
No. 30. *Storm sewage investigations.* September, 1965
No. 31. *Reclamation of water from sewage effluents.* December, 1965
No. 32. *Formation of sulphide in sewers.* March, 1966
No. 33. *Information service on toxicity and biodegradability.* June, 1966
No. 34. *Synthetic detergents* II: *Non-ionic detergents.* September, 1966
No. 35. *Sludge dewatering on drying beds.* December, 1966
No. 36. *Residues of organo-chlorine pesticides in surface waters.* March, 1967
No. 37. *Sponsored investigations with particular reference to the treatment and disposal of trade effluents.* June, 1967
No. 38. *Filter pressing of raw undigested sewage sludges.* September, 1967
No. 39. *Pollution of inland waters by oil.* December, 1967
No. 40. *The use of plastic filter media for biological filtration.* March, 1968
No. 41. *Eutrophication of Inland Waters.* June, 1968
No. 42. *Synthetic detergents* III: *the position in 1968.* September, 1968
No. 43. *Protozoa in sewage-treatment processes.* December, 1968
No. 44. *Simple methods for testing sewage effluents.* March, 1969
No. 45. *Some effects of pollution on fish* II. June, 1969
No. 46. *Disposal of sewage from coastal towns.* September, 1969
No. 47. *Sewage fungus in rivers.* December, 1969

APPENDIX 2

WEIGHT

OUNCES→KILOGRAMMES→GRAMMES

Ounces (oz)	Kilogrammes (kg)	Grammes (g)
1	0·02835	28·35
2	0·05670	56·70
3	0·08505	85·05
4	0·11340	113·40
5	0·14175	141·75
6	0·17010	170·10
7	0·19845	198·45
8	0·22680	226·80
9	0·25515	255·15
10	0·28350	283·50
11	0·31185	311·85
12	0·34020	340·20
13	0·36855	368·55
14	0·39690	396·90
15	0·42525	425·25
16	0·45360	453·60

KILOGRAMMES→GRAMMES→OUNCES

Kilogrammes (kg)	Grammes (g)	Ounces (oz)
1	1 000	35·28
2	2 000	70·56
3	3 000	105·84
4	4 000	141·12
5	5 000	176·40
6	6 000	211·68
7	7 000	246·96
8	8 000	282·24
9	9 000	317·52
10	10 000	352·80

British: Metric Units—Conversion Tables

POUNDS → KILOGRAMMES

Pounds (lb)	Kilogrammes (kg)
1	0·4536
2	0·9072
3	1·3608
4	1·8144
5	2·2680
6	2·7216
7	3·1752
8	3·6288
9	4·0824
10	4·5360

KILOGRAMMES → POUNDS

Kilogrammes (kg)	Pounds (lb)
1	2·205
2	4·410
3	6·615
4	8·820
5	11·025
6	13·230
7	15·435
8	17·640
9	19·845
10	22·050

1 tonne = 1 000 kg = 0·984207 U.K. tons

LENGTH

INCHES → MILLIMETRES → METRES

Inches (in)	Millimetres (mm)	Metres (m)
1	25·4	0·0254
2	50·8	0·0508
3	76·2	0·0762
4	101·6	0·1016
5	127·0	0·1270
6	152·4	0·1524
7	177·8	0·1778
8	203·2	0·2032
9	228·6	0·2286
10	254·0	0·2540
11	279·4	0·2794
12	304·8	0·3048

MILLIMETRES → METRES → INCHES

Millimetres (mm)	Metres (m)	Inches (in)
1	0·001	0·03937
2	0·002	0·07874
3	0·003	0·11811
4	0·004	0·15748
5	0·005	0·19685
6	0·006	0·23622
7	0·007	0·27559
8	0·008	0·31496
9	0·009	0·35433
10	0·010	0·39370

FEET → METRES

Feet (ft)	Metres (m)
1	0·3048
2	0·6096
3	0·9144
4	1·2192
5	1·5240
6	1·8288
7	2·1336
8	2·4384
9	2·7432
10	3·0480

METRES → FEET

Metres (m)	Feet (ft)
1	3·281
2	6·562
3	9·843
4	13·124
5	16·405
6	19·686
7	22·967
8	26·248
9	29·529
10	32·810

YARDS → METRES

Yards (yd)	Metres (m)
1	0·9144
2	1·8288
3	2·7432
4	3·6576
5	4·5720
6	5·4864
7	6·4008
8	7·3152
9	8·2296
10	9·1440

METRES → YARDS

Metres (m)	Yards (yd)
1	1·094
2	2·188
3	3·282
4	4·376
5	5·470
6	6·564
7	7·658
8	8·752
9	9·846
10	10·940

MILES → KILOMETRES → METRES   KILOMETRES → METRES → MILES

Miles	Kilometres (km)	Metres (m)	Kilometres (km)	Metres (m)	Miles
1	1·609	1 609	1	1 000	0·6214
2	3·218	3 218	2	2 000	1·2428
3	4·827	4 827	3	3 000	1·8642
4	6·436	6 436	4	4 000	2·4856
5	8·045	8 045	5	5 000	3·1070
6	9·654	9 654	6	6 000	3·7284
7	11·263	11 263	7	7 000	4·3498
8	12·872	12 872	8	8 000	4·9712
9	14·481	14 481	9	9 000	5·5926
10	16·090	16 090	10	10 000	6·2140

AREA

SQ. INCHES → SQ. MILLIMETRES → SQ. METRES   SQ. METRES → SQ. MILLIMETRES → SQ. INCHES

Sq. inches	Sq. millimetres (mm^2)	Sq. metres (m^2)	Sq. metres (m^2)	Sq. millimetres (mm^2)	Sq. inches (in^2)
1	645·2	0·0006	1	1 000 000	1 550
2	1290·4	0·0013	2	2 000 000	3 100
3	1935·6	0·0019	3	3 000 000	4 650
4	2580·8	0·0026	4	4 000 000	6 200
5	3226·0	0·0032	5	5 000 000	7 750
6	3871·2	0·0039	6	6 000 000	9 300
7	4516·4	0·0045	7	7 000 000	10 850
8	5161·6	0·0052	8	8 000 000	12 400
9	5806·8	0·0058	9	9 000 000	13 950
10	6452·0	0·0065	10	10 000 000	15 500
11	7097·2	0·0071			
12	7742·4	0·0077			

SQ. FEET → SQ. METRES

Sq. feet (ft²)	Sq. metres (m²)
1	0·0929
2	0·1858
3	0·2787
4	0·3716
5	0·4645
6	0·5574
7	0·6503
8	0·7432
9	0·8361
10	0·9290

SQ. METRES → SQ. FEET

Sq. metres (m²)	Sq. feet (ft²)
1	10·76
2	21·52
3	32·28
4	43·04
5	53·80
6	64·56
7	75·32
8	86·08
9	96·84
10	107·60

SQ. YARDS → SQ. METRES

Sq. yards (yd²)	Sq. metres (m²)
1	0·8361
2	1·6722
3	2·5083
4	3·3444
5	4·1805
6	5·0166
7	5·8527
8	6·6888
9	7·5249
10	8·3610

SQ. METRES → SQ. YARDS

Sq. metres (m²)	Sq. yards (yd²)
1	1·196
2	2·392
3	3·588
4	4·784
5	5·980
6	7·176
7	8·372
8	9·568
9	10·764
10	11·960

ACRES → HECTARES → SQ. METRES

Acres	Hectares (ha)	Sq. metres (m^2)
1	0·4047	4 047
2	0·8094	8 094
3	1·2141	12 141
4	1·6188	16 188
5	2·0235	20 235
6	2·4282	24 282
7	2·8329	28 329
8	3·2376	32 376
9	3·6423	36 423
10	4·0470	40 470

HECTARES → SQ. METRES → ACRES

Hectares (ha)	Sq. metres (m^2)	Acres
1	10 000	2·471
2	20 000	4·942
3	30 000	7·413
4	40 000	9·884
5	50 000	12·355
6	60 000	14·826
7	70 000	17·297
8	80 000	19·768
9	90 000	22·239
10	100 000	24·710

SQ. MILES → SQ. KILOMETRES → HECTARES

Sq. miles	Sq. kilometres (km^2)	Hectares (ha)
1	2·589	258·88
2	5·178	517·76
3	7·766	776·64
4	10·355	1035·52
5	12·944	1294·40
6	15·533	1553·29
7	18·122	1812·17
8	20·710	2071·05
9	23·299	2329·93
10	25·888	2588·81

SQ. KILOMETRES → HECTARES → SQ. MILES

Sq. kilometres (km^2)	Hectares (ha)	Sq. miles
1	100	0·386
2	200	0·772
3	300	1·158
4	400	1·545
5	500	1·931
6	600	2·317
7	700	2·703
8	800	3·089
9	900	3·475
10	1 000	3·861

VOLUME

CUBIC FEET → CUBIC METRES → LITRES

Cubic feet (ft^3)	Cubic metres (m^3)	Litres (l)
1	0·02832	28·32
2	0·05664	56·64
3	0·08496	84·96
4	0·11328	113·28
5	0·14160	141·60
6	0·16992	169·92
7	0·19824	198·24
8	0·22656	226·56
9	0·25488	254·88
10	0·28320	283·20

CUBIC METRES → LITRES → CUBIC FEET

Cubic metres (m^3)	Litres (l)	Cubic feet (ft^3)
1	1 000	35·31
2	2 000	70·62
3	3 000	105·93
4	4 000	141·24
5	5 000	176·55
6	6 000	211·86
7	7 000	247·17
8	8 000	282·48
9	9 000	317·79
10	10 000	353·10

CUBIC YARDS → CUBIC METRES → LITRES

Cubic yards (yd^3)	Cubic metres (m^3)	Litres (l)
1	0·7646	764·6
2	1·5292	1 529·2
3	2·2938	2 293·8
4	3·0584	3 058·4
5	3·8230	3 823·0
6	4·5876	4 587·6
7	5·3522	5 352·2
8	6·1168	6 116·8
9	6·8814	6 881·4
10	7·6460	7 646·0

CUBIC METRES → LITRES → CUBIC YARDS

Cubic metres (m^3)	Litres (l)	Cubic yards (yd^3)
1	1 000	1·308
2	2 000	2·616
3	3 000	3·924
4	4 000	5·232
5	5 000	6·540
6	6 000	7·848
7	7 000	9·156
8	8 000	10·464
9	9 000	11·772
10	10 000	13·080

British: Metric Units—Conversion Tables

GALLONS → CUBIC METRES → LITRES

Gallons (gal)	Cubic metres (m^3)	Litres (l)
1	0·00455	4·546
2	0·00909	9·092
3	0·01364	13·638
4	0·01818	18·184
5	0·02273	22·730
6	0·02728	27·276
7	0·03182	31·822
8	0·03637	36·368
9	0·04091	40·914
10	0·04546	45·460

CUBIC METRES → LITRES → GALLONS

Cubic Metres (m^3)	Litres (l)	Gallons (gal)
1	1 000	220
2	2 000	440
3	3 000	660
4	4 000	880
5	5 000	1 100
6	6 000	1 320
7	7 000	1 540
8	8 000	1 760
9	9 000	1 980
10	10 000	2 200

GALLONS → CUBIC METRES → MEGALITRES

Gallons (gal)	Cubic metres (m^3)	Megalitres (Ml)
1 000 000	4 546	4·546
2 000 000	9 092	9·092
3 000 000	13 638	13·638
4 000 000	18 184	18·184
5 000 000	22 730	22·730
6 000 000	27 276	27·276
7 000 000	31 822	31·822
8 000 000	36 368	36·368
9 000 000	40 914	40·914
10 000 000	45 460	45·460

LINEAR VELOCITY

FEET/SEC → METRES/SEC

Feet/sec (ft/sec)	Metres/sec (m/sec)
1	0·3048
2	0·6096
3	0·9144
4	1·2192
5	1·5240
6	1·8288
7	2·1336
8	2·4384
9	2·7432
10	3·0480

METRES/SEC → FEET/SEC

Metres/sec (m/sec)	Feet/sec (ft/sec)
1	3·281
2	6·562
3	9·843
4	13·124
5	16·405
6	19·686
7	22·967
8	26·248
9	29·529
10	32·810

FEET/MINUTE → METRES/SEC

Feet/minute (ft/min)	Metres/sec (m/sec)
1	0·0051
2	0·0102
3	0·0152
4	0·0203
5	0·0254
6	0·0305
7	0·0356
8	0·0407
9	0·0457
10	0·0508

METRES/SEC → FEET/MINUTE

Metres/sec (m/sec)	Feet/minute (ft/min)
1	196·86
2	393·72
3	590·58
4	787·44
5	984·30
6	1181·16
7	1378·02
8	1574·88
9	1771·74
10	1968·60

RATES OF FLOW

CUBIC FEET/SEC → CUBIC METRES/SEC
→ LITRES/SEC

Cubic feet/sec (ft³/sec)	Cubic metres/sec (m³/sec)	Litres/sec (l/sec)
1	0·02832	28·32
2	0·05664	56·64
3	0·08496	84·96
4	0·11328	113·28
5	0·14160	141·60
6	0·16992	169·92
7	0·19824	198·24
8	0·22656	226·56
9	0·25488	254·88
10	0·28320	283·20

CUBIC METRES/SEC → LITRES/SEC
→ CUBIC FEET/SEC

Cubic metres/sec (m³/sec)	Litres/sec (l/sec)	Cubic feet/sec (ft³/sec)
1	1 000	35·31
2	2 000	70·62
3	3 000	105·93
4	4 000	141·24
5	5 000	176·55
6	6 000	211·86
7	7 000	247·17
8	8 000	282.48
9	9 000	317·79
10	10 000	353·10

CUBIC FEET/MINUTE → LITRES/SEC
→ CUBIC METRES/SEC

Cubic feet/minute (ft³/min)	Litres/sec (l/sec)	Cubic metres/sec (m³/sec)
1	0·47	0·00047
2	0·94	0·00094
3	1·41	0·00141
4	1·88	0·00188
5	2·35	0·00235
6	2·82	0·00282
7	3·29	0·00329
8	3·76	0·00376
9	4·23	0·00423
10	4·70	0·00470

CUBIC METRES/SEC → LITRES/SEC
→ CUBIC FEET/MINUTE

Cubic metres/sec (m³/sec)	Litres/sec (l/sec)	Cubic feet/minute (ft³/min)
1	1 000	2118·6
2	2 000	4237·2
3	3 000	6355·8
4	4 000	8474·4
5	5 000	10 593·0
6	6 000	12 711·6
7	7 000	14 830·2
8	8 000	16 948·8
9	9 000	19 067·4
10	10 000	21 186·0

SURFACE LOADING

CUBIC FEET/SQ. FOOT → LITRES/SQ. METRE → CUBIC METRES/SQ. METRE

Cubic feet/ sq. foot (ft³/ft²)	Litres/ sq. metre (l/m²)	Cubic metres/sq. metre (m³/m²)
1	304·8	0·3048
2	609·6	0·6096
3	914·4	0·9144
4	1 219·2	1·2192
5	1 524·0	1·5240
6	1 828·8	1·8288
7	2 133·6	2·1336
8	2 438·4	2·4384
9	2 743·2	2·7432
10	3 048·0	3·0480

CUBIC METRES/SQ. METRE → LITRES/SQ. METRE → CUBIC FEET/SQ. FOOT

Cubic metres/sq. metre (m³/m²)	Litres/ sq. metre (l/m²)	Cubic feet/ sq. foot (ft³/ft²)
1	1 000	3·281
2	2 000	6·562
3	3 000	9·843
4	4 000	13·124
5	5 000	16·405
6	6 000	19·686
7	7 000	22·967
8	8 000	26·248
9	9 000	29·529
10	10 000	32·810

GALLONS/SQ FOOT → LITRES/SQ METRE → CUBIC METRES/SQ METRE

Gallons/ sq. foot (gal/ft²)	Litres/sq. metre (l/m²)	Cubic metres/ sq. metre (m³/m²)
1	48·93	0·0489
2	97·86	0·0979
3	146·79	0·1468
4	195·72	0·1957
5	244·65	0·2446
6	293·58	0·2936
7	342·51	0·3425
8	391·44	0·3914
9	440·37	0·4404
10	489·30	0·4893

CUBIC METRES/SQ METRE → LITRES/SQ METRE → GALLONS/SQ FOOT

Cubic metres/ sq. metre (m³/m²)	Litres/sq. metre (l/m²)	Gallons/ sq. foot (gal/ft²)
1	1 000	20·44
2	2 000	40·88
3	3 000	61·32
4	4 000	81·76
5	5 000	102·20
6	6 000	122·64
7	7 000	143·08
8	8 000	163·52
9	9 000	183·96
10	10 000	204·40

VOLUMETRIC LOADING

GALLONS/CUBIC YARD → LITRES/CUBIC METRE → CUBIC METRES/CUBIC METRE

CUBIC METRES/CUBIC METRE → LITRES/CUBIC METRE → GALLONS/CUBIC YARD

Gallons/ cubic yard (gal/yd³)	Litres/ cubic metre (l/m³)	Cubic metres/ cubic metre (m³/m³)	Cubic metres/ cubic metre (m³/m³)	Litres/ cubic metre (l/m³)	Gallons/ cubic yard (gal/yd³)
1	5·946	0·0059	1	1000	168·2
2	11·892	0·0119	2	2000	336·4
3	17·838	0·0178	3	3000	504·6
4	23·784	0·0238	4	4000	672·8
5	29·730	0·0297	5	5000	841·0
6	35·676	0·0357	6	6000	1 009·2
7	41·622	0·0416	7	7000	1 177·4
8	47·568	0·0476	8	8000	1 345·6
9	53·514	0·0535	9	9000	1 513·8
10	59·460	0·0595	10	10 000	1 682·0

POUNDS/CUBIC YARD → KILOGRAMMES/CUBIC METRE

KILOGRAMMES/CUBIC METRE → POUNDS/CUBIC YARD

Pounds/Cubic yard (lb/yd³)	Kilogrammes/ Cubic metre (kg/m³)	Kilogrammes/ Cubic metre (kg/m³)	Pounds/Cubic yard (lb/yd³)
1	0·5933	1	1·6856
2	1·1867	2	3·3712
3	1·7800	3	5·0568
4	2·3733	4	6·7424
5	2·9666	5	8·4281
6	3·5600	6	10·1137
7	4·1533	7	11·7993
8	4·7466	8	13·4849
9	5·3400	9	15·1705
10	5·9333	10	16·8561

HEAT

British Thermal Units (B.Th.U.'s)	Kilojoules (KJ)
1	1·055
2	2·110
3	3·165
4	4·220
5	5·275
6	6·330
7	7·385
8	8·440
9	9·495
10	10·550

BRITISH THERMAL UNITS → KILOJOULES

Kilojoules (KJ)	British Thermal Units (B.Th.U.'s)
1	0·9478
2	1·8956
3	2·8434
4	3·7912
5	4·7390
6	5·6868
7	6·6346
8	7·5824
9	8·5302
10	9·4780

KILOJOULES → BRITISH THERMAL UNITS

KILOGRAMMES OF B.O.D.

VOLUME (CUBIC METRES)

B.O.D. (mg/l)	1	2	3	4	5	6	7	8	9	10	20	30	40	50	60	70	80	90
1	0·001	0·002	0·003	0·004	0·005	0·006	0·007	0·008	0·009	0·01	0·02	0·03	0·04	0·05	0·06	0·07	0·08	0·09
2	0·002	0·004	0·006	0·008	0·01	0·012	0·014	0·016	0·018	0·02	0·04	0·06	0·08	0·10	0·12	0·14	0·16	0·18
3	0·003	0·006	0·009	0·012	0·015	0·018	0·021	0·024	0·027	0·03	0·06	0·09	0·12	0·15	0·18	0·21	0·24	0·27
4	0·004	0·008	0·012	0·016	0·020	0·024	0·028	0·032	0·036	0·04	0·08	0·12	0·16	0·20	0·24	0·28	0·32	0·36
5	0·005	0·010	0·015	0·020	0·025	0·030	0·035	0·040	0·045	0·05	0·10	0·15	0·20	0·25	0·30	0·35	0·40	0·45
6	0·006	0·012	0·018	0·024	0·030	0·036	0·042	0·048	0·054	0·06	0·12	0·18	0·24	0·30	0·36	0·42	0·48	0·54
7	0·007	0·014	0·021	0·028	0·035	0·042	0·049	0·056	0·063	0·07	0·14	0·21	0·28	0·35	0·42	0·49	0·56	0·63
8	0·008	0·016	0·024	0·032	0·040	0·048	0·056	0·064	0·072	0·08	0·16	0·24	0·32	0·40	0·48	0·56	0·64	0·72
9	0·009	0·018	0·027	0·036	0·045	0·054	0·063	0·072	0·081	0·09	0·18	0·27	0·36	0·45	0·54	0·63	0·72	0·81
10	0·01	0·02	0·03	0·04	0·05	0·06	0·07	0·08	0·09	0·1	0·2	0·3	0·4	0·5	0·6	0·7	0·8	0·9
20	0·02	0·04	0·06	0·08	0·1	0·12	0·14	0·16	0·18	0·2	0·4	0·6	0·8	1·0	1·2	1·4	1·6	1·8
30	0·03	0·06	0·09	0·12	0·15	0·18	0·21	0·24	0·27	0·3	0·6	0·9	1·2	1·5	1·8	2·1	2·4	2·7
40	0·04	0·08	0·12	0·16	0·20	0·24	0·28	0·32	0·36	0·4	0·8	1·2	1·6	2·0	2·4	2·8	3·2	3·6
50	0·05	0·10	0·15	0·20	0·25	0·30	0·35	0·40	0·45	0·5	1·0	1·5	2·0	2·5	3·0	3·5	4·0	4·5
60	0·06	0·12	0·18	0·24	0·30	0·36	0·42	0·48	0·54	0·6	1·2	1·8	2·4	3·0	3·6	4·2	4·8	5·4
70	0·07	0·14	0·21	0·28	0·35	0·42	0·49	0·56	0·63	0·7	1·4	2·1	2·8	3·5	4·2	4·9	5·6	6·3
80	0·08	0·16	0·24	0·32	0·40	0·48	0·56	0·64	0·72	0·8	1·6	2·4	3·2	4·0	4·8	5·6	6·4	7·2
90	0·09	0·18	0·27	0·36	0·45	0·54	0·63	0·72	0·81	0·9	1·8	2·7	3·6	4·5	5·4	6·3	7·2	8·1
100	0·1	0·2	0·3	0·4	0·5	0·6	0·7	0·8	0·9	1·0	2·0	3·0	4·0	5·0	6·0	7·0	8·0	9·0
200	0·2	0·4	0·6	0·8	1·0	1·2	1·4	1·6	1·8	2·0	4·0	6·0	8·0	10·0	12·0	14·0	16·0	18·0
300	0·3	0·6	0·9	1·2	1·5	1·8	2·1	2·4	2·7	3·0	6·0	9·0	12·0	15·0	18·0	21·0	24·0	27·0
400	0·4	0·8	1·2	1·6	2·0	2·4	2·8	3·2	3·6	4·0	8·0	12·0	16·0	20·0	24·0	28·0	32·0	36·0
500	0·5	1·0	1·5	2·0	2·5	3·0	3·5	4·0	4·5	5·0	10·0	15·0	20·0	25·0	30·0	35·0	40·0	45·0
600	0·6	1·2	1·8	2·4	3·0	3·6	4·2	4·8	5·4	6·0	12·0	18·0	24·0	30·0	36·0	42·0	48·0	54·0
700	0·7	1·4	2·1	2·8	3·5	4·2	4·9	5·6	6·3	7·0	14·0	21·0	28·0	35·0	42·0	49·0	56·0	63·0
800	0·8	1·6	2·4	3·2	4·0	4·8	5·6	6·4	7·2	8·0	16·0	24·0	32·0	40·0	48·0	56·0	64·0	72·0
900	0·9	1·8	2·7	3·6	4·5	5·4	6·3	7·2	8·1	9·0	18·0	27·0	36·0	45·0	54·0	63·0	72·0	81·0
1 000	1·0	2·0	3·0	4·0	5·0	6·0	7·0	8·0	9·0	10·0	20·0	30·0	40·0	50·0	60·0	70·0	80·0	90·0

246

KILOGRAMMES OF B.O.D.

VOLUME (CUBIC METRES)

B.O.D. (mg/l)	100	200	300	400	500	600	700	800	900	1 000	2 000	3 000	4 000	5 000	6 000	7 000	8 000	9 000
1	0·1	0·2	0·3	0·4	0·5	0·6	0·7	0·8	0·9	1·0	2·0	3·0	4·0	5·0	6·0	7·0	8·0	9·0
2	0·2	0·4	0·6	0·8	1·0	1·2	1·4	1·6	1·8	2·0	4·0	6·0	8·0	10·0	12·0	14·0	16·0	18·0
3	0·3	0·6	0·9	1·2	1·5	1·8	2·1	2·4	2·7	3·0	6·0	9·0	12·0	15·0	18·0	21·0	24·0	27·0
4	0·4	0·8	1·2	1·6	2·0	2·4	2·8	3·2	3·6	4·0	8·0	12·0	16·0	20·0	24·0	28·0	32·0	36·0
5	0·5	1·0	1·5	2·0	2·5	3·0	3·5	4·0	4·5	5·0	10·0	15·0	20·0	25·0	30·0	35·0	40·0	45·0
6	0·6	1·2	1·8	2·4	3·0	3·6	4·2	4·8	5·4	6·0	12·0	18·0	24·0	30·0	36·0	42·0	48·0	54·0
7	0·7	1·4	2·1	2·8	3·5	4·2	4·9	5·6	6·3	7·0	14·0	21·0	28·0	35·0	42·0	49·0	56·0	63·0
8	0·8	1·6	2·4	3·2	4·0	4·8	5·6	6·4	7·2	8·0	16·0	24·0	32·0	40·0	48·0	56·0	64·0	72·0
9	0·9	1·8	2·7	3·6	4·5	5·4	6·3	7·2	8·1	9·0	18·0	27·0	36·0	45·0	54·0	63·0	72·0	81·0
10	1·0	2·0	3·0	4·0	5·0	6·0	7·0	8·0	9·0	10·0	20·0	30·0	40·0	50·0	60·0	70·0	80·0	90·0
20	2·0	4·0	6·0	8·0	10·0	12·0	14·0	16·0	18·0	20·0	40·0	60·0	80·0	100·0	120·0	140·0	160·0	180·0
30	3·0	6·0	9·0	12·0	15·0	18·0	21·0	24·0	27·0	30·0	60·0	90·0	120·0	150·0	180·0	210·0	240·0	270·0
40	4·0	8·0	12·0	16·0	20·0	24·0	28·0	32·0	36·0	40·0	80·0	120·0	160·0	200·0	240·0	280·0	320·0	360·0
50	5·0	10·0	15·0	20·0	25·0	30·0	35·0	40·0	45·0	50·0	100·0	150·0	200·0	250·0	300·0	350·0	400·0	450·0
60	6·0	12·0	18·0	24·0	30·0	36·0	42·0	48·0	54·0	60·0	120·0	180·0	240·0	300·0	360·0	420·0	480·0	540·0
70	7·0	14·0	21·0	28·0	35·0	42·0	49·0	56·0	63·0	70·0	140·0	210·0	280·0	350·0	420·0	490·0	560·0	630·0
80	8·0	16·0	24·0	32·0	40·0	48·0	56·0	64·0	72·0	80·0	160·0	240·0	320·0	400·0	480·0	560·0	640·0	720·0
90	9·0	18·0	27·0	36·0	45·0	54·0	63·0	72·0	81·0	90·0	180·0	270·0	360·0	450·0	540·0	630·0	720·0	810·0
100	10·0	20·0	30·0	40·0	50·0	60·0	70·0	80·0	90·0	100·0	200·0	300·0	400·0	500·0	600·0	700·0	800·0	900·0
200	20·0	40·0	60·0	80·0	100·0	120·0	140·0	160·0	180·0	200·0	400·0	600·0	800·0	1 000·0	1 200·0	1 400·0	1 600·0	1 800·0
300	30·0	60·0	90·0	120·0	150·0	180·0	210·0	240·0	270·0	300·0	600·0	900·0	1 200·0	1 500·0	1 800·0	2 100·0	2 400·0	2 700·0
400	40·0	80·0	120·0	160·0	200·0	240·0	280·0	320·0	360·0	400·0	800·0	1 200·0	1 600·0	2 000·0	2 400·0	2 800·0	3 200·0	3 600·0
500	50·0	100·0	150·0	200·0	250·0	300·0	350·0	400·0	450·0	500·0	1 000·0	1 500·0	2 000·0	2 500·0	3 000·0	3 500·0	4 000·0	4 500·0
600	60·0	120·0	180·0	240·0	300·0	360·0	420·0	480·0	540·0	600·0	1 200·0	1 800·0	2 400·0	3 000·0	3 600·0	4 200·0	4 800·0	5 400·0
700	70·0	140·0	210·0	280·0	350·0	420·0	490·0	560·0	630·0	700·0	1 400·0	2 100·0	2 800·0	3 500·0	4 200·0	4 900·0	5 600·0	6 300·0
800	80·0	160·0	240·0	320·0	400·0	480·0	560·0	640·0	720·0	800·0	1 600·0	2 400·0	3 200·0	4 000·0	4 800·0	5 600·0	6 400·0	7 200·0
900	90·0	180·0	270·0	360·0	450·0	540·0	630·0	720·0	810·0	900·0	1 800·0	2 700·0	3 600·0	4 500·0	5 400·0	6 300·0	7 200·0	8 100·0
1 000	100·0	200·0	300·0	400·0	500·0	600·0	700·0	800·0	900·0	1 000·0	2 000·0	3 000·0	4 000·0	5 000·0	6 000·0	7 000·0	8 000·0	9 000·0

INDEX

Acidity 27, 158, 177
Acids 167
  as coagulants 75
Activated sludge 89, 121, 130
  activity determination 100
  densification 133, 210
  growth of organisms 98–100
  microscopic examination 100
  'quality' or 'condition' tests 100–106
  settleability 103
Activated sludge process 13, 26–27, 60, 80, 89–106, 110, 113, 118, 133, 157, 168, 189, 197, 201
  aeration, modified forms of 95–98
  analyses at various stages of purification 34
  coarse bubble aeration 207–208
  comparison with percolating filters 106
  control of 19
  diffused air system 90–91
  efficiency improvements 207
  flow diagram 95
  high-rate 97
  mechanical aeration systems 91–98
  nitrification 104–106
  oxygen requirements 96
  plant design memorandum 190
  plant design trends 207–208
  stages involved 94–95
Aerated lagoons 107
Aeration, coarse bubble 207–208
  extended 97–98
  INKA process 91, 207
  mechanical systems 91–98, 207
  modified forms of 95–98
  step 95, 97
  tapered 96

Aeration cone 73
Aeration plants 191
Aeration tanks 90, 94–95, 97
Aerobic biological treatment 79–108, 180, 201
Aerobic digestion 97, 100
Aerobic processes 13
Agricultural trade wastes 156
Air diffusers 90
Albuminoid nitrogen 26, 35
Algae 107
Algal blooms 117
Algal growths 202
Algal lagoon 118
Alkalinity 27, 158, 177
Allyl thiourea 162, 211
Aluminium chloride 125
Aluminium chlorohydrate 125
Aluminium hydroxide 74
Aluminium salts 125
Aluminium sulphate 74, 125
Alumino-ferric 74, 159, 173
Alundum sludge 90
Ammonia 26
  conversion to nitrites and nitrates 14
Ammoniacal nitrogen 116, 117
Ammonium bromide 212
Amoebae 102
Anaerobic bacteria 131
Anaerobic bacterial action 77
Anaerobic decomposition 18
Anaerobic digestion 180
Anaerobic processes 15–16
Anaerobic retting 177
Analysis of sewage 16–28
  of sewage and sludge 32–37
Andover Borough Council 207
Anisopus fenestralis 85

Anodising wastes 178
Appeals 7, 8, 154, 155
Area, conversion tables 235–237
Arsenic 162
Atomised suspension technique 209
Atritor 141
Auckland 107
Australia 107
Auto Analyser 19
Auto-digestion 100
Automatic control 198
Automation 198, 208

Bacteria 9, 18, 107
 aerobic 13
 anaerobic 15, 131
 methane-forming 129
 nitrate-reducing 89
 removal of 113
Baffle board 64
Baffle plates 149
Baffles 64
Balancing tanks 197
Banks clarifier 115
Barrhead 127
Barston Works 116
Beaches, sewage pollution of 201
Beating process 171
Beckton Works 214
Bio-aeration system 94
Biochemical oxidation 226
Biochemical Oxygen Demand. See B.O.D.
Bio-flocculation 73
Biological film 112, 121
Biological oxidation 187
Biological processes 16
Biological purification 80
Biological treatment, aerobic 201
 methods 211–212
Birmingham 182–183
Bleaching 165, 211
Bletchley Sewage Works 140
Boby-Imacti Immedium Filter 114
B.O.D. (Biochemical Oxygen Demand) 4, 5, 14, 30, 31, 157, 181
 calculation of 15
 kilogrammes of (tables) 245–246
 long-term 30
B.O.D./P.V. ratios 23–24
B.O.D./sludge loading factor 190
B.O.D. test 19, 22–24
Bournemouth 201

Bradford 106
Brewery wastes 175, 212
British Columbia 212
British units, conversion of 227–228
Bromine-82 212
Buchner funnel test 139
Bulking 103–104
Bury Corporation Sewage Works 117
Buxton Sewage Works 136
Chlorophyll 107
Chromium 161
Canning industry wastes 179
Carbon, organic, determination of 24
Carbonisation of coal 168
Case hardening 178
Cell growth 98–99
Cellulose fibre 157, 171
Cess-pools 186–187
Channels, open, flow measurement in 146–150
Chemical analysis. See Analysis
Chemical coagulation. See Coagulation
Chemical conditioners 124
Chemical manufacture wastes 177
Chicago 209
Chloride 27
Chlorinated copperas 74, 75, 125
Chlorination 13, 58–59, 186, 199, 201
Chlorine 58–59, 112, 125, 167, 202

Clarifier, Banks 115
 upward-flow gravel bed 185
Clariflocculator 71
Classifier 57
Clean Rivers (Estuaries and Tidal Waters) Act (1960) 7, 8
Coactor 127
Coagulants 73–76, 125, 139, 180
Coagulation 73–76, 80, 121, 157, 173
Coal, carbonisation of 168
Coastal towns 200–202
Coke ovens 169
Colloidal matter 71, 113
Combined nitrogen 26
Comminutor 48, 52, 201
Composting 123–124
Concentrated Ammoniacal Liquor Plant 168
Concrete, sewer, corrosion of 12
Conditioners 127, 136
Consolidation tanks 122
Contact beds 79
Contact stabilisation plant 191

Index

Conversion of units 227
Conversion tables 232–246
Cooling water 194–195
Copper 162
Copperas 74, 125, 136, 173
Corrosion of sewer concrete 12
Cost, of trade waste treatment 181
  of water reclamation 196
Cotton bleaching, dyeing, printing and finishing wastes 165–168, 211
Crawley 110
Crude ammoniacal liquor 168
Curie (unit) 164
Cyanides 160
Cyclone separator 141

Dairy wastes 176
Davyhulme Sewage Works 76–77
DDT 85
Denitrification 89, 210
Deoxygenation, rate of 14
Deoxygenation constant 15
Detention time 62, 212
Detergents 78, 91, 131, 162–163, 213–218
  anionic 213, 216, 217
  biodegradability test 217
  cationic 214
  hard 128, 216, 217
  non-ionic 213, 217
  soft 216, 217, 218
Detritus 53
Detritus tanks 46, 54
Dewatering of sludge. *See* Sludge
Dichromate value (C.O.D.) test 24
Diffused air system 90–91
Diffusers, air 90
  dome 90, 91
  flat-plate 91
Digestion, anaerobic 180
  causes of failure 131
  high-rate anaerobic 128
  mesophilic 128
  of organic matter 130, 224
  thermophilic 128
Digestion plant 133
Digestion processes, sludge 128–133
Digestion tanks 130
Dilution 109
Discharge rate, for flume 149
  for Venturi meter 151
Disintegrator 48
Distilling industry wastes 175

Distributors 81, 188, 206
Dome diffusers 90, 91
Domestic sewage 186
Donaldson sludge density index (S.D.I.) 102
Dorr Clariflocculator 71
Dorr Detritor 56
Dorr-Oliver Fluo-solids (FS) system 209
Drene 215
Dry-weather flow (D.W.F.) 41–47
  calculation of average 41
Drying, sludge 140–141, 209
Drying beds, sludge 133–135
Dunstable 116
D.W.F. *See* Dry-weather flow
Dyeing and dyes 167

Effluent, sewage. *See* Sewage effluent
  trade. *See* Trade wastes
Electrolytic treatment of sewage 202
Electro-plating wastes 178
Elutriation 126
Emergency service 205
Endogenous respiration 99–100
Engineering process wastes 179
Equivalent weight 221–222
Eutrophication 117, 202
Extended aeration 97–98

Farm wastes 180
Fellmongering wastes 174
Fermentation 129, 130
Ferric chloride 74, 75, 125
Ferric hydroxide 74
Ferric sulphate 125
Ferrous bicarbonate 74
Ferrous hydroxide 74
Ferrous sulphate 74, 75
Fertilisation 202
Fertilisers, sewage effluents as 195
  sludge as 120, 123, 124, 127, 135, 140, 142
Film, biological 112, 121
Film growth 84–85
Filter pressing 135–136
Filters, Banks clarifier 115
  enclosed aerated 86
  nitrifying 117
  percolating 13, 60, 79–89, 105, 121, 188, 197
    alternating double 85, 87, 207
    area of media required 225

Filters percolating *continued*
   comparison with activated sludge process 106
   design trends 206
   high-rate 85–89
   low-rate 89
   media 80–81
   series operation 84, 85
   two-stage 84, 207
   roughing 81
   sand 112, 113–115, 200
      Immedium 114
      upward flow 114–115
   slow sand 185
   sprinkling. *See* Filters, percolating
   storm 45
   trickling. *See* Filters, percolating
   vacuum 126, 136–139
   see also Filtration
Filtration, high-rate 85, 88
   intermittent downward 79
   vacuum 126, 136–139
   see also Filters
Finishing wastes 167
Fire hazards 158, 204
Flax retting wastes 176
Flocculating agents 73–76
Flocculating aids 126
Flocculating chamber 71
Flocculation 76
   bio 73
   mechanical 71, 111, 157
   mechanism 73
'Floor' synthetic medium 81
Flotation 133
Flow in sewers 38
   see also Dry-weather flow
Flow measurement 145–152
   automatic recording 146
   in open channels 146–150
   in pipes 150–152
   points at which measurements are required 145
   radioactive tracers 212
Flow rates, conversion tables 241
Flue gas 158
Fluidised bed 142
Flumes 149–150
Fly nuisance 85, 88, 106
Foaming 91, 98
Food industry wastes 183
Formaldehyde 161
Føyn process 202

Francis's Formula for flow through weirs 147, 148
Freezing 127
Fuel 36, 80, 104, 141
Furnaces, multiple hearth 141

Gammexane 85
Gas, sludge 36–37, 80, 104, 140
Gas liquor 32
Gas works wastes 169
Germany 97, 209
Grass plots 191
Grease and its removal 48, 57–58, 136, 159
Great Ryburgh 212
Grit chambers, aerated 57
Grit disposal 57
Grit removal 53–57
Grit tanks 54–57
Grit washer 57
Guernsey 203

Halifax 127, 141, 206
Hard water 10
Hazelwood Lane Sewage Plant 112
Heat exchanger 127
Heat treatment 128
Heat units, conversion tables 244
Highways, spillages on 204
Horticultural trade wastes 156
Housing and Local Government, Minister of 7, 8, 46, 154, 155, 156
Housing and Local Government, Ministry of 110, 187
Huddersfield 127, 141
Humus 13
Humus sludge 83, 121, 130
Humus tanks 83, 85, 89, 111, 116, 118, 121, 188
Hydrated lime 73
Hydraulic control 46
Hydrogen ion concentration 27
Hydrogen peroxide 211
Hydrogen sulphide 2, 11–12, 199
Hypochlorite 58, 167, 186

Imhoff cone 25
Imhoff tanks 15, 77
Immedium sand filter 114
Incineration and incinerators 52, 142, 211

Index

Incubator tests 28
Industrial wastes. *See* Trade wastes
Infiltration water 38, 44
Inflammable liquids 158, 204
INKA Aeration Process 91, 207
Insecticides 85, 177
Insects 107
  see also Fly nuisance
Institute of Sewage Purification 202
Iron 162
Iron salts 74–75, 125
Irrigation 79, 114, 116, 195
  sub-surface 189
  surface 189

Joint Sewerage Board 194

Kessener Brush 93, 212
Kessener plants 207
Kier liquor 165, 170
Komline-Sanderson coil filter 137

Lagoons 107, 116, 118, 123
Lakes, clarification 115
Land disposal 122, 142, 181
Land plots 123
Land treatment 13, 79, 116–118, 189
Laundry trade wastes 154, 156
Lead 162
Leap weir 43
Leather manufacture wastes 173–174
Legislation 7–8, 153–156, 164, 180, 193, 211, 212
Length, conversion tables 233–235
Lime 13, 118, 125, 136, 173
Linear velocity, conversion tables 240
Load variations in sewage works 197
Loading, surface, conversion tables 242
  volumetric, conversion tables 243
Local authorities 110, 153–156, 181, 184, 205, 211
Local Government Board 41
Los Angeles 118
Luton Sewage Works 111, 113, 200, 216

Macerator 52, 201, 203
Macro-organisms 121
Magnesium hydroxide 202
Magnetic flow meter 152
Malthouse wastes 212
Maple Lodge Works 216
Maturation ponds 116

McGowan strength 30, 31, 182, 183
Mechanisation 198, 208
Mendia process 202, 203
Mercerising wastes 167
Mercury 162
Mersey, River 195
Mesophilic digestion 128
Metal industries wastes 177–179, 183
Metals, toxic 161
Methane 18, 36
Methane-forming bacteria 129
Methylene blue 28
Micro-organisms 115
Micro-respirometer 100
Microscopic examination of activated sludge 100
Micro-strainers 111–113, 200
Mieder type sludge scraper mechanism 66
Mineral matter 36
Minister of Housing and Local Government 7, 8, 46, 154, 155, 156
Ministry of Housing and Local Government 110, 186
Ministry of Technology 28
Mixed liquor 90, 91, 95
Mogden Formula 182
  modified form of 183
Mogden Sewage Works 54, 182, 199, 210
Mohlman sludge volume index (S.V.I.) 102
Mosquitoes 107

Nappe 149
Neutralisation 158, 167
New York 97
New Zealand 107
Nickel 161
Nitrate-reducing bacteria 89
Nitrates 26, 79, 117
Nitrification 14, 15, 23, 27, 80, 86, 89, 104–106
Nitrates 26
Nitrogen 118
  albuminoid 26, 35
  ammoniacal 116, 117
  combined 26
Nitrosification 14
Nordell number 100
Normal solution 220
Nutrients, removal of 117–118

Odeeometer 100

Odours 2, 10–12, 18, 58, 77, 79, 85, 88, 106, 107, 121, 127, 128, 158, 199, 202
Oil and its removal 48, 57–58, 159, 179
Organic compounds 89, 118, 131, 162
Organic matter, anaerobic breakdown of 15–16
   digestion of 130, 224
   tests for presence of 19
Organic polymers 126
Organic substances in sewage 10
Organisms, activated sludge, growth of 98–100
Orifice 46
Orifice meter 151
Overflows, storm sewage 205
   storm water 40, 41–44
   designs of 43–44
   setting of 43
Oxidation 14, 18, 80, 89
   biochemical 226
   biological 188
   total 97
Oxidation ditches 191
Oxidation ponds 106–107, 191
Oxygen 157
   dissolved 2, 3, 4, 16, 18, 100, 115, 202
   uptake of 14
Oxygen absorption, from potassium dichromate 24
   from potassium permanganate 21
Oxygen solubility and temperature 2
Oxygenation capacity 91, 207
Ozone 160, 195

Paper and board manufacture 170–171
Pasveer oxidation ditch 168, 190, 212
Paxman Roto-plug concentrator 140
Pentachlorophenol 131, 162, 211
Perborates 162
Per cent purification 220
Percolating filters 188
Permanganate value 30, 31, 157
Permanganate value test 19, 21
Pesticides 177
pH 27, 35, 129
   in coagulation 73–75
pH meter 28
Phelps' law 14, 226
Phenols 161
Phosphates 117, 118, 202
Photosynthesis 15
Phytoplankton 117

Pickling liquors 178
Pipes, flow measurement in 150–152
Plant growth 36
Pollution 4, 41, 193–219
Polyelectrolytes as coagulants 76
Polyvinyl chloride 81, 136
Ponding 84, 85, 88
Ponds, maturation 116
   oxidation 106–107, 191
Population equivalent 181
   British 32
   USA 31
Porteous process 127
Portsmouth 203
Potassium dichromate, oxygen absorption from 24
Potassium permanganate, oxygen absorption from 21
Poultry evisceration wastes 212
Pre-aeration 76–77
Preliminary treatment 48–59
Printing wastes 167
Protozoa 9, 100, 102, 107
Pseudo-colloidal matter 71
Psychoda alternata 85
Psychoda severini 85
Public Health Act (1961) 154, 180, 211
Public Health (Drainage of Trade Premises) Act (1937) 153
Pulp, manufacture 169–170
   paper and board manufacture wastes 169–172
Putrefaction 18

Radioactive isotopes 163
Radioactive materials 162, 212
Radioactive Substances Act (1960) 164, 212
Radioactive tracers 152
Radioactivity 163–165, 212
Radio-isotopes 212
Raymond system of sludge drying 141
Re-aeration 110
Recirculation 83, 86, 89
Recirculation ratio 86
Reclamation processes 194–197
Redox potential 18
Reduction 18
Regional authorities 193–194
Regionalisation of sewage treatment 193
Respirometer 14
Retention time, measurement of 152
Retting, anaerobic 177

Index 253

Retting liquors 177
Rhizopods 102
River Authorities 7, 8, 110, 191, 200
River Mersey 195
River pollution 41
River Purification Authorities 8
River Stour 201
Rivers (Prevention of Pollution) Acts 7, 8, 180, 193
Rivers (Prevention of Pollution) (Scotland) Acts 8
Road tankers 204
Rotifers 102
Royal Commission 1, 4, 5, 7, 19, 22, 25, 30, 34, 41, 44, 45, 62, 77, 79, 84, 85, 93, 97, 98, 106, 109, 110, 116, 117, 191, 199, 206, 207
Rye Meads Sewage Works 110, 113

Safety precautions, sludge digestion plants 37
Salford 77, 80, 85
Sampling 20–21
Scenedesmus 118
Scotland 8, 127
Scott drier 141
Scouring organisms 84
Scrapers 70
   flight-conveyor 67
   link-belt 67
   Mieder type 66
Screening chamber 49
Screenings, disposal of 52
Screens and screening 44, 46, 47–52, 84, 157, 201
   coarse 49
   drum 50
   fine 49
   fixed 49
   grab type 51
   hand-raked 49
   mechanically raked 50
   moving 49, 50
   raking mechanisms 50
   size of 49
   sonic 209
   wire rope 50
Scum-boards 47, 64, 70
Scum removal 64
Sea disposal of sludge 122, 200–202
Sedimentation 157, 173, 187, 201
   aids to 71–73
   primary 60–78

Sedimentation *continued*
   secondary 67
   theoretical basis of 61
   two-stage 63
Sedimentation tanks 44, 45, 48, 62–71, 190, 197
   continuous-flow 61–63
   design of inlets and outlets 63
   design trends 206
   fill-and-draw quiescent 65
   horizontal-flow 65, 77
      manually desludged 66
      mechanically desludged 66
   operational performance 70
   overflow rate 71
   parallel operation 63
   primary 60, 63, 120, 206
   radial-flow 69
   secondary 63
   series operation 63
   size of 62
   sludge removal 64
   surface loading 70
   types of 65–71
   upward-flow 67–68
Seed sludge 129
Seeding 18, 98
Self-purification 2, 110
Separan-2610 126
Separation weir 44–45
Septic condition 4, 58
Septic sewage 11, 58
Septic tanks 15, 77–78, 187
Settleable solids test 25
Settling tanks 25
Severn River Board 116
Sewage, acid 12
   alkaline 12
   analysis of 16–28
   analysis of average strength domestic, at various stages of purification 35
   analysis of weak, average and strong 32
   chlorination of 13
   dilution 4
   domestic 38, 187
   fresh 10
   nature of 9–15
   objectionable character of 10
   organic substances in 10
   septic 11, 58
   solid matter in 9
   stabilisation 14

254

Sewage *continued*
   stale 10
   strength test 23
   treatability of 28–29
   untreated 2
   water content of 9
Sewage disposal, Royal Commission on. *See* Royal Commission
Sewage effluent, B.O.D. 5
   disposal of 191–192, 200–202
   methods of improving final 109–119
   new discharge of 6–7
   reclamation of water from 194
   Royal Commission standard. *See* Royal commission
   stability of 28
   uses 194
Sewage pollution of beaches 201
Sewage sludge. *See* Sludge
Sewage strength 28–32
   assessment of 30
   factors influencing 29
Sewage treatment, aerobic processes 13–15
   anaerobic processes 17–18
   biological processes 18
   electrolytic 202
   industrial, analyses at various stages of purification 32–34
   regionalisation of 193
   tertiary 107, 111, 116, 195
Sewage treatment plants, small 186–192
Sewage works, appearance of 198–199
   automation 198, 208
   capacity of 5, 19, 28
   choice of site 6
   control of 19
   layout 198–199
   load variations in 197
   processes in 6
Sewer concrete, corrosion of 12
Sewerage systems 38–41
   combined 39–41, 205
   partially separate 39–41
   separate 39–41
Sewers, flow in 38
Sheffield 136, 142
Sheffield Bio-aeration system 94
SI units, conversion of 227–228
Silver 162
Simcar aerator 93
Simplex plants 91, 207
Skimming tanks 58
Slaughterhouse wastes 175

Sloughing (or unloading) 84, 121
Slow sand filters 191
Sludge, activated. *See* Activated sludge
   activity determination 100
   age 102
   analysis 19
   analysis of commoner types 35–36
   as fertiliser 120, 123, 124, 127, 135, 140, 142
   humus 121, 130
   liquid, disposal of 122–124
   moisture content 121–122
   primary 120
   secondary 127
   seed 129
   specific resistance of 137–139
   types of 120–121
   volume of 121–122
   weight-volume-moisture-solids relationships 223
Sludge cake 122, 209
Sludge density index 102, 103
Sludge disposal 64, 66, 70, 76, 77, 78, 83, 95, 98, 107, 120–144, 187, 198
   burning 142
   composting 123–124
   future problems 210
   land disposal 122, 142
   liquid sludge 122–124
   methods and uses 141
   sea disposal 122
   trends in 209–211
Sludge drying 133–135, 140–141, 209
Sludge gas 36–37, 80, 104, 140
Sludge growth index 98
Sludge indexes 102, 103
Sludge removal. *See* Sludge disposal
Sludge tips 141
Sludge treatment 120–144
   atomised suspension technique 209
   dewatering methods 133–140
      chemical conditioning 124–126
      drying beds 133–135
      filter pressing 135–136
      Paxman Roto-plug concentrator 140
      sonic screens 209
      treatment to facilitate 124–133
      vacuum filtration 136–139
   digestion 128–133
      safety precautions 37
   digestion tanks 15
   Dorr-Oliver fluo-solids (FS) system 209
   drying methods 140–141, 209

Index

Sludge *continued*
 elutriation 126
 freezing 127
 future problems 210
 heat treatment 128
 Porteous process 127
 trends in 208–211
 Zimmermann wet combustion process 209
Sludge volume index 102, 103
Soakaway system 191
Sodium chlorite 211
Sodium Dobane PT sulphonate 216
Sodium tetrapropylene benzene sulphonate 215-216
Solid matter, in sewage 9
 removal of 48
 settleable, test for 25
 suspended 60, 61, 62, 71, 111, 157
 determination of 25
 total, test for 25
Sonic screens 209
South Africa 113, 116, 195
Specific resistance of sludge 137–139
Spent liquor 168–169
Sphaerotilus 104
Spillages on highways 204
Standard solutions 220
Step aeration 95, 97
Stevenage 195
Stilling-chamber overflow 44
Stokes' Law 61
Storm sewage overflows 205
Storm water overflows 40, 41–44
 designs of 43–44
 setting of 43
Storm water treatment 44–47
Stour, River 201
Strength of sewage. *See* Sewage strength
Sulphides 159, 167, 172
Sulphuretted hydrogen. *See* Hydrogen sulphide
Sulphuric acid 75, 125
Supernatant water, removal of 133–134
Surface loading, conversion tables 242
Suspended solids, determination of 25
Synthetic resins 81
Syphons 82

Tanks, aeration 90, 94–95, 97
 balancing 197
 consolidation 122
 detritus 46, 54

Tanks *continued*
 digestion 130
 grit 54–57
 humus 83, 85, 89, 111, 116, 118, 121, 189
 Imhoff 15, 77
 sedimentation. *See* Sedimentation tanks
 septic 15, 77–78, 188
 settling 25
 skimming 58
 sludge digestion 15
Tanning liquors 173–174
Tapered aeration 96
Tar distillation 169
Technology, Ministry of 28
Teepol 163, 215
Temperature and oxygen solubility 2
Templewood-Hawksley Sludgemaster 210
Tertiary treatment 107, 111, 116, 195
Tests, sewage analysis 16–28
 simple methods 28
 see also under specific tests
Thames Conservancy Board 112
Thermophilic digestion 128
Throttle-pipe 44, 46
Tidal waters, sewage disposal in 122, 200–202
Tip drainage 203
Tipping tray 188
Total oxidation 97
Total solids test 25
Toxic metals 161
Toxicity of trade wastes 29
Trade wastes 38, 44, 76, 81, 89, 93, 106, 109, 111, 120, 153–185, 197
 Act of 1937 on 153
 agricultural 156
 brewery 175, 212
 bye-laws 154, 155
 canning industry 179
 characteristic figures 165
 characteristics of 156–165
 charges for treatment at sewage works 181–185
 chemical manufacture 177
 coal carbonisation 168
 conversion to equivalent of sewage 31
 cotton bleaching, dyeing, printing and finishing 165–168, 211
 dairies 176
 distilling industry 175
 engineering processes 179

Trade *continued*
  farm wastes 180
  fellmongering 174
  flax retting 176
  food industry 183
  high alkalinity or acidity 158
  high oil and grease content 159
  high oxygen demand 157
  high suspended solids content 157
  high-temperature 158
  horticultural 156
  inflammable liquids 158
  influence on strength of sewage 29
  injurious and inhibitory constituents 159
  laundry 154, 156
  leather manufacture 173–174
  legislation. *See* Legislation
  metal industries 177–179, 183
  new discharge of 7
  population equivalent 31
  poultry evisceration 212
  pulp, paper and board manufacture 169–172
  radioactive 163–165, 212
  slaughterhouses 175
  strength of 30
  toxicity of 29
  trends in 211–212
  types of 165–181
  viscose rayon manufacture wastes 171–172
  wool manufacture 58, 75, 76, 106, 120, 125, 159, 172
Tree planting in sewage works 198
Trenching 122

Ultimate Oxygen Demand 24, 30
Ultrasonic waves 73
Ultra-violet lamps 112
United States 60, 76, 88, 97, 107, 111, 118, 125, 126, 141, 198, 209
Units, conversion of 227
Upper Tame Main Drainage Authority 116
Upward-flow gravel bed clarifiers 191
Urine 10

Vacuum filtration 126, 136–139
Van Kleeck formula 130, 224
Velocity, linear, conversion tables 240
Venturi meter 150

Viruses 9
Viscose rayon manufacture wastes 171–172
Volatile liquids 158, 167
Volume, conversion tables 238–239
Volumetric loading, conversion tables 243

Wandle Valley Sewage Works 125
Warburg apparatus 100
Warrington 195
Wash waters 166
Water, hard 10
Water consumption 29
Water content of sewage 9
Water pollution control trends 193–219
Water Pollution Research Laboratory 195, 217
Water Resources Act (1963) 8
Water supply 110, 117, 194
Wedge Wire Filter Bed 135
Weedkillers 177
Weight, conversion tables 232
Weir 64, 82, 146–149
  compound 148
  fully contracted 147
  notched 206
  rectangular 147
  separation 44–45
  sharp-edged 146
  side 43, 45, 46
  suppressed 147
  V-notch 146
West Hertfordshire Main Drainage Authority 125, 137, 141, 143, 216–217
West Kent Sewerage Board 141
Witham 116
Wool removal from sheepskins 174
Wool-scouring wastes 58, 75, 76, 106, 120, 125, 159, 172
Wuppertal-Buchenhofen Sewage Works 97, 209

Zimmermann wet combustion process 209
Zinc 162